TWELVE DAYS OF
TERROR

A thresher shark on display. One of the many 1916 "Man-eaters."

TWELVE DAYS OF
TERROR

A Definitive Investigation of the
1916 New Jersey Shark Attacks

RICHARD G. FERNICOLA, M.D.

The Lyons Press
Guilford, CT
An Imprint of Globe Pequot Press

The Lyons Press is an imprint of The Globe Pequot Press.

Designed by Compset, Inc.

Printed in the United States of America

10 9 8 7 6 5 4 3 2 1

Library of Congress Cataloging-in-Publication Data is available on file.

photo credits
Frontispiece: Courtesy Wayne T. Bell
page xiv: Courtesy George H. Moss, Jr. Archives
page xxxviii: Courtesy John Bailey Lloyd
page 66: Courtesy Clark C. Wolverton
page 138: Courtesy Department Library Services, American Museum of Natural History
page 178: Courtesy Department Library Services, American Museum of Natural History
page 278: Reprinted with permission from "Distribution of the White Shark, Carcharodon Carcharias in the Western North Atlantic" by John G. Casey and Harold L. Pratt, Jr., 1985
Illustrations on pages 202, 204, 205, 209, and 211 by Philip Shaheen III and Danielle Cather.

To my mother

"... and let them have dominion over the fish of the sea ..."

—Genesis, 1:26

Contents

Acknowledgments

During the investigation of a dramatic, historic, and scientifically compelling event, a great number of sources must contribute to the comprehensive success of the project. The creation of this book, therefore, represents a tribute to the long list of generous people and organizations that provided essential pieces to a very complex puzzle.

The thanks I am obligated to bestow likely begins with an anonymous figure I met on a New Jersey-bound train from Washington, D.C., in the early 1980s. I was on my way home for a weekend away from college when a man sitting next to me learned that I was interested in marine biology. The man bought me a beer and described the attacks of 1916. It was the first time I'd heard them in such detail. From there, I consulted with friend, neighbor, and New Jersey waterman Dr. Dennis Sternberg, and the reality of the 1916 attacks began to snowball. The library and microfilm then became my home and works like *Shadows in the Sea* and Richard Ellis' *Book of Sharks* told me that others knew of the New Jersey attacks as well.

The recollections of witnesses who welcomed me into their homes will never be forgotten or underappreciated, and it is from their memories that I derived the driving force and inspiration that made every step of the study a work of intrigue. Posthumous thanks must go to the witnesses George "Red" Burlew, Bill Burlew, Johnson Cartan, Mildred Fisher, John Applegate, Russell Cable, Leroy Smith, Dr. Robert Patterson, and Alva Allen. Thanks must also go to the following witnesses and the relatives of primary witnesses such as Mary Bailey, Jackie Ott, Marion Smith, Annette Baker, Jerry Hourihan, Sarah Ellison, June Rounds, the Vineyard family, and Gwen Reddy. Doris Enterline of Matawan was an absolute angel in her assistance to find many of these sources.

The scientific research and formal assistance of countless individuals also became essential to assembling a coherent and powerful approach to the elusive mystery. Richard Ellis, George Burgess, Peter Klimley, Stewart Springer, John McCosker, Jack Casey, Harold Pratt, Ralph Collier, Marie Levine, and many other premier shark researchers are responsible for shedding light on the debate encapsulated in this volume.

The private collections, photographs, unique research material, and vital contacts that the following men, woman, and organizations provided makes this presentation a complete picture of the period and personalities studied. They are Clark C. Wolverton, Malcolm Barhenburg, Walter Jones, George Moss, James Foley, Patricia Colrick, Tim McMahon, Maurice Cuocci, Wayne T. Bell, Robert Stewart, Dick Labonte, Zoe Wells, Helen Henderson, John Bailey Lloyd, The National Archives, The American Museum of Natural History, The Brooklyn Museum, The University of Pennsylvania Alumni Office, The Smithsonian Institute, The Florida Museum of Natural History, The Ocean County Historical Society, The Spring Lake Historical Society, The Matawan Historical Society, the Matawan Public Library, The Rutgers University Alexander Memorial Library, The Asbury Park

Public Library, Monmouth University Guggenheim Memorial Library, and the Neptune Museum.

Personal thanks must also go to Gregory Fernicola, Robert Fernicola, Roy Fernicola, my cousin, Paul Fernicola, and Joseph Hornick. Philip Shaheen III, Danielle Cather, and Lorraine Shaheen, were indispensible toward the completion of the artwork. I can never forget the innumerable occasions of help and patience provided by Vera Fernicola.

It may not be widely known, but New Jersey is home to some very prominent names in the world of sharks. Marie Levine of Princeton is the director of the Shark Attack Research Institute and specializes in South African attacks. David Doubilet of Elberon is a world-renowned shark photographer whose photos have graced the front pages of *National Geographic* and other noted magazines. Accomplished underwater photographer Stan Waterman also resides in New Jersey, I believe. Peter Benchley, too, is a resident of Princeton, New Jersey. Perhaps it is only fitting that the great effort put forward in this work is an examination of a New Jersey–based story. The subject matter just happens to have worldwide implications.

Bathers on the Jersey Shore at the turn of the twentieth century.

Introduction

In the essential order of prey/predator food chain, there exists a delegation of creatures perched at the highest level with few natural enemies. The polar bear and the Nile crocodile are some notable apex predators that not only rule their respective alimentary niches, but also have been responsible for more than one human death. Biologists and animal behavior specialists do debate the assertion that these creatures are actually seeking out man as prey, but no one will dispute the very real danger that these animals can present to a person caught in the wrong place at the wrong time.

Of all the predators implicated in man-eating events, none conjures up more intense dread than the shark. It is certainly terrifying to be confronted by a bear deep in the woods, or a tiger in a remote village in India, but it is, perhaps, of some small comfort to know that a single shot from a rifle can thwart the danger from such a visible terrestrial predator. It is quite another issue, however, to imagine being a shipwrecked sailor, a surfer, or a beach bather about to be attacked, dismembered, and consumed by a dark, black-eyed monster with razor-sharp teeth, vicelike jaws, and sandpaper-like skin.

What is so terribly disconcerting about this scenario is that the ocean is an environment that hinders both our ability to see the approaching danger and complicates our most natural and effective means of escape. The shark, on the other hand, is the master predator of its domain. Conditions that are only marginal for human comfort and survival are the same conditions that allow this powerful creature to spring suddenly and mercilessly upon us.

In 1975, the release of the major motion picture *Jaws* touched off an international shark attack panic. Peter Benchley, author of the novel, said in an interview commemorating the twenty-fifth anniversary of this classic work, "Completely, inadvertently, it tapped into a very, very deep fear." The Benchley story depicts a fearsome beast of the deep that strikes first at an unsuspecting nighttime skinny-dipper. Next, with an explosion of air-borne blood, the *Jaws* monster devours a pathetically innocent boy riding on a raft and even ventures into uncharted territory to dismember a swimmer recreating in a narrow inlet. As the novel and the film progress, a rampant shark hunt does little to quell the snowballing fear of a seagoing phantom. It's impossible to forget the image of the black-veiled mother mourning the death of her son or the frantic shark hunters who scramble to safety as their dock refuge is pulverized by the twenty-foot white shark. In the end, hopes and tension ride on the skills of the crusty Captain Quint (Robert Shaw) and his two-man crew, Amity police chief Martin Brody (Roy Scheider) and dedicated ichthyologist Hooper (Richard Dreyfuss). The audience's fear of sharks, however, did not end with the roll of screen credits, and to some extent, they live on, almost three decades later, in the minds of those of us who spend time among the breaking waves. The exact origins of this seemingly inherent fear are unknown. What is known, however, is that the mechanical shark attacking the makeshift Hollywood set is neither the first nor nearly the most terrifying instance of shark attack the world has ever seen. What is also known is the fact that world-famous shark hunter Frank

Mundus had predecessors working his dangerous occupation during the early part of the twentieth century.

In reality, many or most of the elements which gave *Jaws* its punch have a correlate in a real-life horror that unfolded eighty-five years ago. Indeed, man's widespread baptism to sharks and attacks took place nearly sixty years before *Jaws* was written, at the beginning of July 1916 when five people were attacked off the New Jersey shore in the course of twelve days. These bizarre attacks were more than enough to send a prosperous East Coast society and the world at large into a climate of shock and confusion over what could now be classified as a once-in-a-millennium event. These events, however, did not take place in a vacuum. They took place during the waning months of an innocent and grand age. War was raging in Europe and tempting U.S. involvement, a polio epidemic was ravaging New York and threatening New Jersey, German U-boats were appearing in American ports, foreign sabotage missions were taunting the northeast, and Pancho Villa was terrorizing U.S. troops in Mexico. During what was expected to be the last golden summer of the pre-war era, President Wilson and several of his cabinet members looked to the placid Jersey Shore for a spectacular respite prior to the presidential elections in November.

As America moved into the twentieth century, it held an image of itself as glorious, confident, and powerful. It was the victory in the Spanish-American War of 1898 that created, for the first time, the reality of the United States as a world power. William McKinley was the country's leader, but Teddy Roosevelt stood as our ideal—an honest, mighty, and progressive giant. By 1914, the nation looked to New Jersey governor and academic scholar Woodrow Wilson to guide it with the safe hand of presidential diplomacy during the early uncertain months of World War I in Europe. "Peace without victory [war]" was Wilson's optimistic motto. Years earlier, Wilson was elected governor of New Jersey on the laurels of his performance as school president of Princeton University. During the second year of Wilson's presidency, however, his intellectual cool was put to the test by several unprecedented assaults. No less than 128 Americans were caught in the deadly and dastardly cross fire in the German torpedo strike on the British liner *Lusitania*. Closer to home, Heinrich Friedrich Albert, an agent for the German government, carelessly left a briefcase behind in a Manhattan elevated railway car. The secret documents in the misplaced briefcase included incriminating propaganda and sabotage plans linked to the German and Austrian embassies. The beginning of 1916 brought the mysterious sinkings of cargo vessels that had just departed from ports including Boston, New York, and Bridgeport. It was later found that the ships were sunk by small "cigar bombs" placed by a notorious German saboteur called the "Dark Invader." During this same period, northeastern U.S. residents were also visited by German U-boats in multiple U.S. ports. The subs, most notably the *Bremen* and *Deutschland,* were cruising the Mid-Atlantic and southern New England coasts and eventually reported a "humanitarian" purpose for their presence. The submarine captains said they desired to transport much needed condensed milk and foodstuffs back to the fatherland to help the babies dying from shortages. But the sabotage blast at Jersey City's Black Tom Island ammunition depot on July 30, 1916, which shattered office windows in Manhattan, made the German presence even more suspect.

By the spring of 1916, the office of the presidency was working in overdrive. It was an election year, and Wilson was committed to a neutrality approach and embraced the slogan, "He kept us out of war." Wilson's competition, on the other hand, painted Wilson as a peace-loving coward whose stance was risking American lives. Wilson responded, not by negative campaigning or uncharacteristic aggression, but by urging a theme of "readiness" and inspiring nationwide Preparedness Day Parades. The May parade in New York was, up to that time, the greatest display of American flags ever seen.

By the summer of 1916, great military drives broke out on assorted fronts. Germany attacked at Verdun while Austria and Italy fought battles along the Isonzo River. The Allies took the offensive along the Somme River in France, the Russians assaulted the Poles in Galatia, and the fleets of Great Britain and Germany clashed off Jutland in the greatest naval battle of the war.

Even though America was not yet engaged in the battles of World War I, the daily headlines gave the nation a peek at the new business of mechanized destruction and the mass killing (often via nerve gas) of young men. In this final summer prior to U.S. involvement in what would become known as the "war to end all wars," the nation was not without military activity at its own borders. In late June, President Wilson ordered 100,000 National Guardsmen to the Mexican border to patrol against attacks from Mexican rebel Pancho Villa. Villa and his soldiers had crossed over into New Mexico and Texas and committed cold-blooded crimes during the previous five months. Now they were being pursued into the heart of Mexico by the troops of legendary general John J. Pershing.

Since the move into the twentieth century and the emergence of the industrial revolution, the world and the United States seemed to be making steady progress on scientific, technological, and manufacturing fronts. While there were only eight thousand registered passenger vehicles in the United States in 1900, the total soared to over a million and a half by 1916. The car at this time was still only a slightly superior replacement for the horse and buggy, and expectations of its performance didn't necessarily mean that it was to speed along the highway. These vehicles were also very difficult to heat. During the winter, motorists would have to fill metal crates with hot coals to place at their feet in order to keep warm. In 1908, the Model T sold for $850. With Henry Ford's mass production techniques, however, he turned out almost two thousand cars a day and the price of a Model T dropped to $360. Dusty, muddy, and hazardous roads were being improved with paving, and the automobile was no longer a mechanical curiosity. President Wilson was the last U.S. president to be driven to his inauguration in a horse-drawn carriage.

Sparked by the war in Europe, the American economy in 1916 reached new and unprecedented heights of prosperity. Such a climate encouraged people like Seattle timber magnate William Boeing to hire twenty-one carpenters and seamstresses to build fifty Model C biplanes for the navy. The use of telephones at this time was becoming more common but still only in affluent households or thriving businesses.

Communication by commercial or military radio was not yet in use, and ships at sea still depended on wireless telegraph transmissions. Telegrams and letters were the main method of private long-distance communication. Telegrams were even used for intracity emergency messages. Newspapers stood as a flourishing medium to provide information to the hungry masses. The print medium of eighty-five years ago, at least in the form of its many daily papers, used a sensationalist writing style. It was as if every city paper was in staunch competition, and the accepted method to deal with such a contest (or combat) was to provide headlines and story "facts" that would outshock a rival. This approach did not necessarily advocate fabrication, but it allowed for a tenuous standard of accuracy when eliciting witness testimony. If the information provided in a story was not riveting in itself, the flowery writing style of the period would more than make up for the mundane subject.

While the world at large was seeing a great many changes, both beneficial and hazardous, medical science was a step behind in its capability to care for new threats on mankind. In the early part of the century, longevity was a major challenge. The swine flu epidemic of 1917, for example, in a world without antibiotics, killed a great many people because of secondary bacterial pneumonia. The World War was being fought without the availability of even basic antibiotics, and successful blood transfusion was only in its infancy. Surgeons were, however, able to perform skin grafting, and antiseptics made the chances of a patient surviving surgery much greater. On the civilian level, there were no ambulance services or any great emergency aid systems. For the unstable patient whose blood pressure was dropping from blood loss or shock, the only hope was to get him to the hospital operating room where saline infusion (saltwater intravenous) and experimental stimulants would be administered. Stimulants such as epinephrine (trade name Adrenalin) were being tried to stabilize a deteriorating trauma victim at this period, and it was also known to help lung congestion in congestive heart failure. In 1916, it was even used intraspinally as an experimental agent to treat the symptoms of infantile paralysis.

Of all the natural plagues of the first half of the 1900s, infantile paralysis, commonly known to us as polio, was one of the most devastating and frightening. Its original name was derived from the misconception that children were its sole victims. Polio, an acute viral infection, presented with fever, headache, stiff neck, and rigid back. Sometimes total paralysis of various muscle groups would be seen. When the respiratory muscles were affected, death was often inevitable. The virus multiplied in the intestines of its host and spread by oral contact with contaminated food or water. Sewage, flies, fly-contaminated food, community pools (pre-dating chlorination), and lakes all played a role in spreading the disease prior to the Salk vaccine of the 1950s. A holdover from precautionary days is the sign at our local pools that states: SHOWER BEFORE ENTERING POOL. Today, epidemiologic studies site poor hygiene and sanitation, city living, and hot summer months as risk factors contributing toward an eruption of widespread outbreaks. One of the worst of such epidemics occurred in 1916. In that fateful summer, New York City was the first area to be stricken by the deadly disease. Hundreds of children were afflicted by the lethal paralysis on a daily basis. Hundreds more attempted to flee the confines of the city to escape what they presumed to be a breeding ground for the deadly germ. When the first few Philadelphia cases were diagnosed during the summer, city residents immediately implicated invading New York children as the cause. So great was the concern about infected children transmitting the disease that guards were stationed at New York /New Jersey borders to keep all persons under the age of sixteen from entering the Garden State unless a medical certificate testified to the fact that they had not been recently exposed to polio patients. In the end, New Jersey recorded 4,055 cases with some counties reporting an infection rate of one in one hundred people.

World political affairs, dilemmas in medical treatment, and industrial advancements dominated much of the scene in the months and years preceding U.S. participation in World War I. Zoology and other natural sciences also saw change in the form of research-based analysis and acquisition of data. The U.S. Life-Saving Service (predecessor to the Coast Guard), for example, would often encounter interesting species of fishes and developed a program of shipping the specimens to assorted museums throughout the country. Amateur naturalists and sportsmen also contributed data from their catches to the growing pool of academic knowledge.

In the 1700s, Swedish naturalist Carolus Linnaeus established a formal system of animal classification. Not only his peers but future generations of biologists used Linnaeus's system as a template to which they added details, facts, and whole species. For a scientific specialist to acquire a particular specimen and scrutinize its anatomy in a laboratory environment was fairly uncomplicated. Other less obvious features—such as maximum length, weight, and, more important, behavior—were sometimes left to conjecture or even to the imagination. Direct observation and description of a mounted specimen is one thing, but the opportunity to analyze the natural behavior and habits of a wild species is a much more complex and difficult task. If a knowledge deficit existed among the naturalists and scientists of 1916, it was not merely secondary to undiscovered lab and field data; it also must have related to the fact that many exotic animals were out of reach in remote or undeveloped regions. It would take years of organized expeditions, financial backing, determination, and even luck to answer many questions that books and exhibits simply could not reveal.

Even today, countless creatures that inhabit the deep sea are true mysteries to all of us. In 1938, fishermen working in the Indian Ocean, off South Africa, discovered a strange metallic-blue fish almost six feet long. The creature was a coelocanth, the descendant of an ancient line of fishes thought to be extinct for seventy million years. It was only within the last twenty years that the bizarre-looking megamouth shark was discovered. (The first megamouth was accidentally snared by a naval anchor in the Pacific in 1976.) It is not surprising then to expect the scientists of 1916 to have a fairly primitive if not inaccurate appraisal of many aspects of the natural history of sharks. In 1915, Mr. J. Ernest Williamson and his brother spent weeks in a submarine chamber (bathysphere) in the West Indies to view sharks under "direct observation." This was an opportunity to study the shark "under natural conditions that seems never to have been granted before to anyone."* The men were particularly interested in taking motion pictures of the shark during feeding. After using assorted baits to attract curiosity in the sharks, the men finally resorted to the carcasses of slaughtered calves "made available by a fortunate chance." The patient pair described how one shark appeared and circled the bait in a very timid fashion. The shark disappeared only to return a short time later with a companion. The process repeated itself over the next six hours until a school of more than thirty sharks was inspecting the quarry. The men then witnessed a snap by a single shark at the hanging bait that triggered a ferocious, all-out attack on the carcasses. The viciousness was so blinding that the sharks were noted to attack one another. What the Williamson brothers observed, for the first time, was the prototypical shark "feeding frenzy." The graphic details of the feeding sequence lends credence to the experiment's authenticity, yet, the observers did estimate some of the sharks to be as long as

---

Scientific American, July 1916.

thirty feet. This suggests at least some exaggeration in their data (unless, of course, their porthole combined with the water acted as a magnifying glass). The men justifiably emphasized the ferocity of the feeding sharks and some of their observations were recorded in London's *Knowledge* magazine and the *New York Times*.

Despite the efforts of the Williamsons and scores of reports from around the globe relating to the tenacity and man-eating potential of large sharks, it was clear that from 1891 to July 1916 the scientific world expressed the firm view that sharks will *not* attack a living man, at least not in temperate waters. The community of scholars who were touted as the premier authorities on shark behavior included a commissioner of fisheries, a museum director, an ichthyologist, and a steamship-line mogul.

The first and most famous gesture to raise public and academic awareness to the risk of a shark attacking a human came from millionaire banker and adventurer Hermann Oelrichs. Oelrichs was certainly an impressive character and was known as much for his intensely likeable personality and mental acuity as he was for his physical prowess. By the age of forty, Oelrichs stood as the New York representative of the North German Lloyd Steamship Company and was rubbing elbows with the likes of *Titanic*'s White Star Line executive, Joseph Bruce Ismay. He was even an instrumental figure in introducing the game of polo to America. Despite his status as one of the richest men in the U.S. and his marriage to a senator's daughter, he would often tempt fate by the dangerous act of swimming out of New York Harbor to greet vessels of his steamship line when they were still miles from entering port. Near the end of his challenging swim, Oelrichs would often lure the vessel toward him, his antics convincing the ship's captain that he was a swimmer in need of assistance. Each time a captain would direct his crew to help Oelrichs aboard, the robust and invincible Oelrichs would refuse the helping hand and demand that he be able return to shore under his own power.

Twenty-five years before the infamous New Jersey attacks, Hermann Oelrichs put the shark question to a practical test by offering, through the columns of the *New York Sun*, a reward of $500 "for an authenticated case of a man having been attacked by a shark in temperate waters [in the U.S., North of Cape Hatteras, North Carolina]." The subject of the true threat of shark bites on human beings was also addressed by articles in *Forest and Stream* in 1897, in the *New York Times* in the summer of 1915, and at other times by U.S. government officials.

Neither the Oelrichs reward nor U.S. government publications elicited authenticated evidence of an attack. There were anecdotal accounts of attacks taking place in Georgia, Florida, Cuba, and the Gulf of Mexico, and the *Times* article in August 1915 did trigger other reports (in the form of letters to the editor) of attacks from the distant past. One letter, from a reader in East Orange, New Jersey, testified to three known attacks in Sydney, Australia. In that correspondence, three boys on separate occasions and in assorted locations of Sydney Harbor were allegedly attacked at a time when it was common practice to dispose of blood and offal from meat factories directly into the harbor. One of the boys involved in the Sydney attacks was said to have been dangling his legs from a dock when his foot was seized, and he was dragged into the water never to be seen again. Another letter writer, from Glen Cove, New York, sent in an account considered very unusual by the mere fact that it occurred in the northeastern U.S. "In 1870," the man wrote, "off Horton's Point on Long Island Sound, a crewman from a schooner that had stalled in the wind, went for a quick swim off the vessel. Fellow shipmates suddenly spotted a shark closing in on him and began to shout warnings as they got out on a lifeboat for a rescue attempt. Before the men got to the crewman, the shark had grabbed his hip. The struggling victim attempted to blind the marauder and, as the boat approached, the shark let go." The injured man was reportedly treated for several weeks in Greenport,

Long Island, and survived the ordeal. "This story," the writer states, "was told to him by a sailor who resided in Greenport."

Unfortunately, most of the early accounts and recollections of shark attack cases were predominately the product of hearsay and were contributed by nonacademic members of the general public. One letter, however, was submitted by none other than Dr. Frederic A. Lucas, director of the American Museum of Natural History. Then recognized as America's foremost expert on sharks, Dr. Lucas responded to the *Times* article with what he considered to be "two fairly reliable references" from excerpts he had earlier received after his work on the *Forest and Stream* article of 1897. The accounts describe a man who lost his leg near a meat-cutting wharf in Bombay and a fatal attack at sea off the Hawaiian Islands. Lucas explained these as situations in which the shark was "provoked" and was inshore responding to obvious stimuli. Despite his contribution of these accounts, in April of 1916, Lucas made it quite clear that he was one of the greatest skeptics of the assertions that sharks are capable of exhibiting man-eating behavior under natural and common circumstances or in temperate waters. Over and over, members of the scientific community would view the threat of shark attack, at least in nontropical waters, a nonissue. Dr. Hugh M. Smith, U.S. Commissioner of Fisheries, and Dr. H. F. Moore, Deputy Commissioner of the Fisheries Service in Philadelphia, each concurred that, except for the possibility of rare attacks in the tropics, shark attacks against man are not recorded in history. Moore mentioned that he knew of one rare case when, in the Fiji Islands, a native woman was bathing in a river some distance from salt water when she lost one of her legs to a shark.

During the early twentieth century, the migratory habits and range locations of sharks were also thought to be understood. James M. Meehan, State Fish Commissioner of Pennsylvania and one-time Philadelphia Aquarium director, felt that only harmless sharks inhabited near-shore waters. He also felt that if a powerful and aggressive pelagic (deep sea) shark were to stray into shallow water, that shark would certainly be a harmless juvenile version. Other shark researchers pro-posed that sharks were commonly great travelers that cover enormous distances in an incredibly short length of time. This view was not nec-essarily incorrect and modern data does support such a view, at least for some species. Such a proposition on speedy, long-distance-swim-ming movement was based not on tagging studies, however, but on a famous case in which documents were recovered in a shark's stomach. In that case, an eighteenth-century American privateer captain in Ja-maican waters was found guilty of documented crimes against the British because of papers he had thrown overboard that very soon after were recovered from a captured shark's stomach off the coast of Haiti. The "shark's papers" were said to have been preserved and ex-hibited at the Institute of Jamaica in Kingston.

In an analysis of the species' specific inhabitancy ranges reported in publications during this period, assorted authorities cited climatic lo-cations somewhat different from what we know of shark populations today. In all fairness to these devoted scientists and naturalists, how-ever, as we shall see, the range/migratory impressions of these early twentieth-century shark mavens, although at odds on several accounts with our current knowledge, may not necessarily have been incorrect, but simply different.

Among the shark species mentioned by the experts as possessing "dangerous potential," were the tiger shark, the blue shark, and the great white (also known as the man-eater). G. Brown Goode, U.S. Fish Commissioner in 1881, states in one of his early reports that from the Chesapeake Bay to New England, between 1820 and 1860, only three specimens of the man-eating (great white) variety were sighted on the Atlantic coast, and their average recorded length was nine feet. Dr. Bashford Dean, head curator at the Smithsonian Institute at that time, did not believe that a white or tiger shark had ever been identified north of Cape Hatteras. As alluded to earlier, the documented maxi-mum lengths reported by some renowned scientists (other than Goode) was often grossly inaccurate as well. Some shark specialists identified the maximum length of the great white to be in the forty-foot range! Today, we know the white shark to reach a length of a little over twenty feet, but only on the rarest of occasions.

Of all the shark lore of the late nineteenth and early twentieth centuries, the topic of unprovoked shark attacks, in which a beast would sever a person's bones or otherwise produce fatal wounds, was viewed with the most academic skepticism. Two curators at Philadelphia's Academy of Natural Sciences, the oldest institution of its kind in America, were among the many to dispute a shark's man-eating prowess. Dr. Henry W. Fowler, a prominent ichthyologist, and Dr. Henry Skinner, an eminent scientist, asserted that no shark could bite off a man's leg with a single bite. The gentlemen even went so far as to declare that sharks do not possess the power in their jaws that would be expected from their enormous size.

One scientist of this period who seems, at least in retrospect, to stand out for the reasonable accuracy of his assertions on great white population ranges and capacity for lethal behavior, was William T. Hornaday, director of the New York Zoological Park. Hornaday cited the range of the white shark population to be that of "the northern Atlantic coast to Australia and to California." He considered the man-eating capability of the white to be far from myth and also believed that one shark-related fatality in northern waters did indeed occur in 1830 when a man-eating shark overturned a fishing boat in Massachusetts and killed the swamped fisherman.

Just two months before the devastating attacks in 1916, there is no doubt that three New York–based individuals distinguished themselves as the resident experts on sharks in temperate waters. John Treadwell Nichols, curator of the Fishes Wing of the American Museum of Natural History, and Robert Cushman Murphy, director of the Brooklyn Museum, collaborated on a *Brooklyn Museum Science Bulletin* (April 24, 1916) where they discussed "Long Island Fauna—The Sharks." In their publication, they provided a meticulous presentation of the anatomy and natural history of the nineteen cited shark species in that region. Drs. Nichols and Murphy also touched on the question of shark attacks. "Few authenticated cases exist of their attacking a living man in the water," the authors wrote. They did give deserved respect to the great white as a potential man-eater, however, but mentioned that the white is a rare fish in all locales and that it is a popular fallacy to call just any large shark a man-eater. Despite their apparent knowledge on the subject of attacks, Nichols and Murphy felt compelled to defer the final opinion on the matter to Dr. Frederic Lucas, director of the American Museum.

> At the request of the writers, Dr. Lucas has very kindly written for this bulletin the subjoined account relating to the status of sharks as man-eaters. Dr. Lucas' long experience, coupled with his repeated critical investigations of 'shark stories' that arise perennially along our seacoast, eminently fit him to write with finality upon the subject so generally misapprehended.

Dr. Lucas used this opportunity to make definitive statements regarding his conclusions on shark attacks and the true danger from sharks. He stated that not a summer would pass without stories coming from New Jersey and Long Island about a large man-eater being caught. In the end, Lucas stated, "the shark turns out to be a harmless, if ugly-looking, sand shark." Lucas did admit that the white shark and the blue shark were truly dangerous species, but reported that no full-grown individual of this variety was ever caught within hundreds of miles of New York.

In reference to the mechanics of an attack, Lucas not only questioned the authenticity of unprovoked attacks but also questioned a shark's physical capability to perform such an act. "There is a great difference between being attacked by a shark and being bitten by one," Lucas wrote. He felt that if a shark were tangled in a net or in the process of active feeding around offal, a person may simply be bitten by pure and understandable accident. Most striking, however, were

Lucas's assertions relating to a shark's jaw power and cutting capacity. Dr. Lucas did not even believe that a thirty-foot shark (it was his assumption that a shark would grow to this length) could directly sever a human bone with one snap and notes that a shark "is not particularly strong in the jaws." In a *Philadelphia Inquirer* article of July 1916, Dr. Fowler agreed with Lucas's view by stating that "it is beyond the power even of the largest Carcharodon to sever the leg of an adult man." Dr. Murphy also supported Lucas's contentions by detailing his disappointment in witnessing a shark struggle in tearing the blubber of a whale carcass as well as that of a dead sea lion. Lucas concluded the discussion by challenging the reader to consider severing a leg of lamb at one stroke during carving and emphasized that the bones of a sheep are much lighter than those of a man.

Part of the influence of Dr. Lucas's firm stance on the issue of shark attack was the fact that in his boyhood, he spent four years at sea. Lucas's father was a New England clipper ship captain and, by the age of six, Lucas had crossed the Atlantic at his father's side and had already joined his dad on two round-the-world voyages before his eighteenth birthday. Over the course of his many childhood seafaring adventures, he never met a man who had been attacked by a shark or even one who knew of such an incident. Additionally, Lucas later disregarded attack accounts as unreliable hearsay or irrelevant to the debate over true unprovoked attacks or at least unrelated to potential attacks in temperate waters. Most importantly, Lucas placed great significance on the fact that the much publicized Oelrichs reward was never claimed during the millionaire's lifetime. For these reasons Dr. Lucas stated that the danger of being attacked by a shark "is infinitely less than that of being struck by lightening and that there is practically **no** [Dr. Lucas's emphasis] danger of an attack from a shark about our coasts." At this point, to appreciate the impending implications of the strong scientific assertions is relatively easy.

The picture of the summer of 1916 would be unfinished, however, if a few words were not devoted toward a depiction of the splendid social setting that dominated the Jersey Shore. While President Wilson preached "neutrality" in an attempt to keep America from the ugliness of World War I, our great grandparents and grandparents would promenade on the wide boardwalks and partake in basic leisure activities that were unknown to the generations before them. The Garden State at this time was a collection of bountiful farms and an industrial dynamo of cities that, to many, existed as a mere corridor in the shadows of New York and Philadelphia. Unlike the second-class billing that much of the north and western regions of the state may have received, the Jersey Shore, on the other hand, was a region spoken of in nothing less than superlatives.

The Jersey Shore is simply the portion of the state that faces the Atlantic Ocean. It extends from the northern part of the Sandy Hook peninsula to the larger peninsula at Cape May. It is flanked on the northern end by the Raritan Bay and on the southern end by the Delaware River. The Delaware River actually cuts the western boundary of the state, essentially making New Jersey a peninsula in itself. With Philadelphia and Baltimore to its south, and New York and Boston to its north, New Jersey has long stood as the gateway between the country's most industrious cities. Let us not forget that shipping on large sailing vessels was the mode of transportation of the time. Cargo and passenger vessels along the Jersey Shore were as common as the cargo truck on the modern highway.

The Garden State's shoreline stretches along the geographic breast of the continent, 127 miles' worth. Its beaches are essentially the divisions between inlets and small rivers and are, on average, raised only a few feet above the tide level. The beaches consist almost exclusively of white sand, and, in the times before development, the western boundaries of the beachfront were made up of cedars, oaks, pines, and grasses, with an assortment of wild fruits, like beach plums, fox grapes, and huckleberries. The many existing freshwater ponds that extend right up to the shoreline are favorite spots for migrating fowl like swans, Canadian geese, and mallards.

Since the Civil War period through the first decades of the twentieth century, the northern Jersey Shore received regular visits from the likes of Mary Todd Lincoln and no fewer than eleven U.S. presidents. Cape May, New Jersey, may have be the country's oldest seaside resort, but Long Branch was the first one that catered to the elite and, for a time, was easily the most celebrated American summer resort. Long Branch's celebrity status was followed closely by Asbury Park and Atlantic City. Long Branch was even referred to as "Summer New York" because every season, Manhattan's fashionable society moved with all its decadent habits to the shore. Ragtime, the syncopated rhythms and energetic melodies of early jazz, were now finding their way from New York City to some of the boardwalk auditoriums and hotels along the Jersey Shore.

Five miles south of Long Branch, Asbury Park was the chosen vacation spot for other wealthy world travelers, and with Victorian suburbs (which still flourish today) like Ocean Grove, Spring Lake, Allenhurst, Elberon, and Deal, the Jersey Shore basked in a grandeur comparable to that of Palm Beach, the French Riviera, the Hamptons, Cape Cod, and Martha's Vineyard. Prior to commercial air travel and air-conditioning, the eight-mile bluff from Asbury to north Long Branch seemed to be the perennial favorite summer hideaway region for the rich and famous.

Diamond Jim Brady, the Guggenheim brothers, Buffalo Bill Cody, Babe Ruth, popular actress Maggie Mitchell, and tenor Enrico Caruso are but a few of the legends who sought out the Shore for home or respite. Let us also not forget that Jackie Gleason's famous television show *Honeymooners* cited their honeymoon location as Asbury Park. In Elberon alone, presidents U. S. Grant and James Garfield owned large summer homes. When President James Garfield realized he was destined to succumb to a progressive infection from an assassin's bullet in September 1882, he asked the White house staff to transport him away from Washington's stifling heat to allow him to enjoy his last days amid the refreshing sea breeze of the Jersey Shore.

In late spring of 1916, President Wilson decided that the summer transfer of the White House executive offices should be to either Asbury Park or Long Branch. By mid-July, Wilson not only went forward with the summer translocation, but he decided that he would regularly visit those two sites. Congressman Thomas Scully, of Deal, New Jersey, received a handwritten communication from Wilson on his decision to come to New Jersey. In the letter to Scully, Wilson wrote, "The only thing I can say is that I will come. New Jersey means too much to me to make any other answer possible." His residence would be the magnificent Shadow Lawn mansion in West Long Branch and the executive headquarters/summer capital would be situated in the Asbury Park Trust Company Building. Conveniently, Wilson's personal secretary, Joseph P. Tumulty, was already a summer resident of Asbury. The spacious summer headquarters occupied the entire fifth floor of the Trust building, which stood at the most conspicuous business corner of the city. In what was turning out to be one of the warmest summers and one of the hottest political campaigns in recent history, the summer capital building would not only house presidential activities, but would also become the center for voting returns for Wilson's reelection in November. The windows of the executive chambers, however, provided Wilson some tranquility as they gave a superb view of the ocean, the lakes, and the wooded countryside.

As an aside, in July 1916, a controversy over two misplaced ballot "X's" in a Glendale, Florida, school bond issue vote went all the way to the Florida Supreme Court, as did various cases rekindled during the tumultuous U.S. Presidential election of 2000.

In early July, William G. McAdoo, secretary of the Treasury, would settle in a few miles south of Asbury at his summer home in lavish Spring Lake. By July 8, he was greeting the northeast's elite at a reception at Spring Lake's spectacular Essex & Sussex Hotel. In attendance at the "who's who" party would be New Jersey governor James E. Fielder, Senator Oliver H. Brown, and the governor's staff physician and surgeon general of the New Jersey National Guard, Colonel

William G. Schauffler. It is no wonder that every major metropolitan newspaper's society columns had references to towns on the Jersey Shore, where the large ornate hotels enjoyed an international reputation for food and service to entice the wealthy.

As the elite enjoyed such seaside luxuries, the rest of the nation was savoring its "age of innocence." Emerging from the Victorian years of the late 1800s, people were now enjoying newfound leisure time. Baseball and boxing were generally considered America's favorite spectator sports, with the "Babe" playing pitcher for the Boston Red Sox in 1916. Monmouth Park Racetrack on the outskirts of Long Branch was and still is a favorite gambling getaway for those who preferred the excitement of the races to that of boardwalk sight-seeing. The shore was particularly attractive because of opportunities for golfing, bicycling, and the horse betting. It was, however, the shore's vibrant ocean and moody shoreline that attracted the most interest. Both the filthy rich and the average Joe did not mind promenading together along the bluff and boardwalk. The enchantment of seaside concerts from John Philip Sousa and Arthur Pryor, rhythmic ragtime piano tunes, and the Victorian holdover of a festive and filling shore dinner all made the Jersey Shore a piece of summer paradise.

The most important attraction of the shore resorts and one of the most popular of all American recreational activities was sea bathing. Since widescale recreational bathing had begun only two decades before, it was fairly new to all and somewhat forbidding for most. The ocean was often ferocious and scary, and few persons truly knew how to swim (e.g., the freestyle crawl). Despite its inherent risks, swimming was something that everyone wanted to do. Guidebooks and pamphlets instructed on what to wear and how to act. "If your teeth are of the kind which do not grow in your mouth," one from the period mentioned, "beware lest a wave knock them out. Now bounce through the surf with a hop, skip, and a jump. Hold your fingers to your ears and your thumbs to your nostrils. Now dance, leap, tumble, swim, kick, float, or make any motion that seems good to you." Of all the leisure-time activities, it must have been the ocean bathing in the cool, sparkling surf that provided the climax for the city traveler.

It was the custom at this time, for instance, for all young men to take a pre-dinner dip. The bathing areas at all popular locations would be replete with poles and an open area of hanging rope to cling to. For bathing, the men would wear the equivalent of black tanktops with swim tights, long, tight shorts that reached to the knee. The ladies would don what looked like thick, black woolen dress/skirts with lower-end ruffles. Others' costumes were twilled flannel, strong and colored brown, blue, or gray. Some styles went a bit overboard with contrasting stripes and ribbons leading a summer cartoonist to write (after the attacks), "If noise will scare off sharks, there are some bathing costumes that should make the wearers quite safe." The bathing attire on the ladies should never be tight and have no ties but just buttons. The suit was also made in a style that would match the standards of a specific locality. The bathers, at least the ladies, did not undertake actual swimming; rather, they would hold on to the ropes to experience the exhilarating blow of the breakers or simply enjoy what was known as "fanny dunking." Up to the late nineteenth century, separate bathing hours for men and women was quite standard. By the early 1900s, women would commonly be accompanied by a male escort before entering the "violent" surf. It was also common for the woman to wear broad-brimmed straw hats to protect both their faces and their locks from coming in contact with the salt water and the intense sun. Most of the ladies, however, would poke their heads into the waves with reckless disregard for the hat or their hair. Some women also wore a closely fitting oil-silk cap and most of them went in without any footwear. A few, however, fearing the effect of shells, crabs, or sharp stones, wore light, thick-soled sandals. A young gentleman would often make an acquaintance in the surf and would offer a young woman his arm just as he would in a square dance. Sometimes parties of ladies and gentlemen would take shape with all of them taking hands and fumbling up and down through the breakers. The merry laughter would be contagious to the point of general hilarity. Seeing the same group of people at a dignified evening affair must have been most awkward after what had transpired in the surf a few hours before.

By 1916, the main manner of etiquette in ocean bathing, besides the coverage of nearly every inch of flesh, was the policy that hotel employees could not share the same bathing area as the guests. Additionally, bathers were generally not allowed to go beyond the sand wearing their bathing attire. Bathhouses along the boardwalks would be fully available for changing, and such houses or huts were usually primitive structures of boards arranged in long rows, sometimes two or three deep and just out of reach of the waves at high tide. In each hut was a pail of clear water to wash the sand from one's feet before dressing.

There was no American institution more popular than the summer resort for surf bathing. It was hard to find a stretch of good bathing beach along the northern Atlantic coast that did not have some type of hotel nearby to cater to the transient summer visitor. When the blistering summer heat encroached, no other place could match the relief of the sparkling shoreline. One did not even have to bathe to enjoy the refreshing breeze. A quick look at any boardwalk photograph of the period reveals that promenading or sitting on the boardwalk required the right attire for the occasion. The well-dressed boardwalk travelers seem almost as overclothed as the female bathers, especially for a ninety-degree July day. Perhaps the parasol umbrellas provided more shade comfort than we would assume.

So the sand and surf of the Jersey Shore held its appeal for all classes. Of all factors, besides the natural attractiveness, that would contribute to the summer convergence of vacationers to the shore, the convenience of rail travel was most alluring. The railroad tracks had been set down even prior to the establishment of many famous shore locations, and, without exaggeration, trains loaded with thousands of daily visitors from New York and Philadelphia would disembark from depots in places like Asbury Park every ten minutes. The rails were also in place for the convenience of Philadelphia travelers who preferred the more serene and pristine setting of South Jersey beaches, like Beach Haven. This then was the blissful merry atmosphere that brought people to the ocean at a rate and volume never before seen by the civilized world. On July 1, 1916, as a carefree lamb is picked for slaughter by a lurking wolf, this age of innocence would come to a gruesome halt.

Beach Haven, New Jersey. Early 1900s.

CHAPTER **1**

Beach Haven: July 1

ess than twenty miles north of the gambling and saltwater taffy capital of Atlantic City lies a tranquil summer resort called Beach Haven. Eighty-five years ago, located on the eighteen-mile stretch of Long Beach Island, Beach Haven was a place to where couples and families escaped for a true vacation, just as they do today. In the 1800s, the town's sole inhabitants were fishermen and whalers, but by the turn of the century, it was well known to outsiders for its appealing pristine waters, picturesque low-lying dunes, and a noncommercial boardwalk. According to John Bailey Lloyd, the noted historian of Long Beach Island, Beach Haven came into being in 1874 as the brainchild of Tuckerton railway founder, Archelaus Ridgeway Pharo. Pharo's daughter actually wanted to name it Beach Heaven. It was, however, literally a haven from hay fever, as its paucity of mainland vegetation fostered a sneeze-free zone. The men who developed Beach Haven envisioned that their resort would be like no other. It was not on the edge of the bustling mainland, as were resorts like Cape May and Long Branch, but almost five miles at sea. As soon as the Tuckerton Railroad was completed in the 1870s, Pharo deeded 670 acres of his land at the southern

1

end of Long Beach Island. The deeded land was divided into lots, and magnificent hotels like the Parry House, the Bay View House, and, later, the Engleside Hotel were constructed. Every weekend during the summer, steam trains arrived packed with beachgoers from southern New Jersey and Pennsylvania. The summer of 1916 was expected to be one of the most profitable and pleasant that Beach Haven had ever seen, and plans for a widened and lengthened boardwalk were to be implemented by summer's end. Two hundred shade trees adorned the town to welcome visitors, and the new express train from Philadelphia shortened the travel time to just under two hours. The Engleside Hotel and the other lavish island resort quarters opened on schedule by the third week of June. By the July Fourth weekend, the busiest of the year, hotels were fully booked. Since the Fourth fell on a Tuesday, vacationers were anticipating an extended fun-filled holiday.

On Saturday, July 1, 1916, the afternoon Number 8 train from West Philadelphia arrived at Long Beach Island. Among the travelers was twenty-five-year-old Charles Epting Vansant. Anyone who knows Philadelphia in the summer knows that its sweltering heat can be more than enough to send a young, tall, handsome, dark-haired, ambitious businessman racing for the invigorating Jersey surf. As one of the most promising graduates of the University of Pennsylvania's class of 1914, and with a bachelor's degree from the College of Arts and Sciences, Vansant had acquired a position with Philadelphia's Folwell Brothers brokerage firm, where his coworkers described him as a man "of unusual promise with an exceptionally winning personality and charm of a manner which brought to him many friends and admirers."* As a Philadelphia native, residing at 4038 Spruce Street, Vansant's college chums knew him as "Charlie" or "Van"; his popularity was further attested to by his role as glee club member, his position as business manager of the *Record* (the school paper), as well as his positions on the business staff of two other prominent school clubs. His athletic abili-

---

*The Evening Bulletin, Philadelphia, July 3, 1916.

ties landed him a slot on the varsity golf team and the junior varsity baseball team. Charles's spiritual side did not suffer because of his athleticism or popularity, for he was also an active congregant of the Walnut Street Presbyterian Church.

Vansant's scholastic achievements were supplemented by his privileged heritage and fortunate natural endowment. The Vansant family lineage distinguished it as one of the oldest family lines in the country. Charles was the only son of prominent Philadelphia nose and throat physician, Dr. Eugene L. Vansant, whose office was located at 1229 Chestnut Street. Dr. Vansant was a direct descendent of Stophel Garettson Vansant, an immigrant from Holland who ventured to New Amsterdam (New York) in 1647. Since 1700, the Vansant clan lived in Pennsylvania, and Charles and his dad were each entitled to membership in the Society for Founders and Patriots of America.

Besides their familiarity with Philadelphia, the Vansant family was also well acquainted with the pleasures of the southern Jersey Shore. In late June 1916, Dr. Vansant, somewhat preoccupied by the growing infantile paralysis epidemic in New York, suggested a family outing away from the city to take advantage of the long Independence Day weekend. On Saturday, July 1, Charles was joined on the train by his dad and two of his three sisters, Eugenia and Louise. Their ultimate destination was the plush Engleside Hotel, replete with seaside tennis courts and a magnificent spired tower.

The Beach Haven express left Camden at 3:35 P.M. and arrived in Beach Haven at a little after 5:00 P.M. Like an elite cruise liner, the large Beach Haven hotels had two dinner sittings that evening, 6:30 and 8:00 sharp. While the thermometer in Philadelphia was close to ninety, the Vansants were now grateful for the comfortable seventy-eight degree F. seaside temperature. After Charles and his family checked into the Engleside, he turned to his sister Louise and suggested a walk to the beach to take in the late-afternoon sun. Vansant also had another motive for the jaunt to the beach, taking the customary pre-dinner dip in the ocean.

While the Vansant family unpacked at the Engleside, Charles dashed off to the boardwalk bathhouses off Centre Street to change into his swim tights. The water temperature was an inviting sixty-eight degrees, just about average for a blistering day in early July. On his walk toward the surf, Vansant befriended a large Chesapeake Bay retriever and waved to his summer acquaintance, lifeguard Alexander Ott. Alex was just about to go off-duty after a long, hot day of guarding the beach, but he wasn't the type to shirk his late-afternoon duties. His eyes remained focused on every swimmer within his bathing area. Ott, athletic and in his mid-twenties, was not only well suited for his position but perhaps even overqualified. In 1910, Alex had been a prestigious member of the American Olympic swim team and had been guarding on the Jersey Shore for several summers.

By this time, Dr. Vansant and Louise arrived beyond the low dune and settled near the lifeguard stand, just as Charles waded into the surf. He frolicked for several minutes with the playful retriever and then he swam beyond the lifelines. Just beyond the breakers, in chest-deep water, Charles turned and began calling for the suddenly timid pooch. The dog was determined to make his exit from the surf, but Vansant continued calling to the dog in hopes that it would follow him.

Up on the beach, bathers ignored Vansant's shouts. At that same moment, however, a handful of people on the shoreline perceived a dark object just beyond the surf. A crowd gathered as the object came into view. What they saw looked like a black fin slicing across the water from the east toward Vansant. They began to shout heatedly to Vansant, who was unable to understand them and continued calling the dog. Suddenly, one shout from Vansant was louder and higher pitched than the rest, and he began struggling to the beach. (At this point, one can't help but visualize the scene from *Jaws* when the black labrador retriever disappeared in the surf during a game of fetch). At about fifty yards from shore, his shouts became shrieks, and he began frantically splashing. His splashes raised the crimson water airborne. Vansant saw his blood becoming one with the sea. He knew his life was departing with it. He made his way to about three and a half feet of water and got to within forty yards of the beach. Perched on his lifeguard stand, Alexander Ott knew immediately that Charles was in grave danger. He darted across the beach toward the pool of bright red blood encircling his friend. Without hesitation and without a thought for his own safety, Ott managed to get the flailing Vansant to waist-deep water. During the struggle to shore, the shark was never far away. Witnesses on the beach and boardwalk later swore they saw the shark affixed to Vansant's thigh during the entire rescue. W. K. Barklie of Broad Street in Philadelphia was on the beach at the time of the attack and later mentioned that the shark did not let Vansant free until its belly scraped the bottom of the sand. Ott finally received assistance amid the breaking waves from local residents John Everton and Sheridan Taylor. The men locked arms and grabbed Ott to create a human chain to hoist Vansant's quickly deadening weight onto the beach. As Charles lay basically motionless near the high-water mark, Louise Vansant stood frozen in horror, mesmerized by the wide stream of draining blood that flowed along the sand past her feet.

During the early years of my 1916 attack investigation, my second move in research protocol, after perusing primary and secondary sources, was always to attempt to contact a witness or a relative of a witness. In reference to the Vansant attack of July 1, I drew blanks all the way around. I tried to find a Vansant family relation in Philadelphia and struck out. I attempted to find John Everton or Sheridan Taylor, but "no cigar," as they say. My most prolonged attempts were at finding anyone connected to Alexander Ott. Realizing that Ott would probably be in his mid-nineties in 1980, I knew that finding him would be a long shot. My calls to assorted Otts in South Jersey, and eventually to every Ott in the state, went completely unrewarded.

In February 1994, I decided to take a much-needed break from the rigors of my residency at NYU Medical Center to spend a nine-day respite in Kuai, Hawaii. A discovery made on the eight-hour airplane ride proved that an investigative historian should never lose hope. As I flipped through that year's *Sports Illustrated* swimsuit issue, I became intrigued by a feature story devoted to great American swimming pools and water shows. I glanced at the photos throughout the story, and I immediately zeroed in on an antique sepia-toned photo of a tall, handsome man in his early forties wearing vintage swimwear and holding an adorable blond kid with a Dutch-boy haircut. The caption beside the photo read, "In the '30s, crowds flocked to the prodigious pool at the Biltmore to see the aquatics of Jackie Ott the Aqua Tot and his father, Alexander." I almost dropped the magazine. I knew immediately that this could be no mere coincidence. This Alexander Ott was in the correct age range, and, of course, the swimming parallel was obvious. No wonder I could not find any trace of Alexander Ott in New Jersey. He had long ago traded in his summertime paradise of South Jersey for the perennial splendor of South Florida.

The *Sports Illustrated* article describes the Coral Gables Biltmore Hotel swimming pool, built in 1926, as the largest pool in Florida and, for many years, the largest hotel pool in the world. Motorboats could even tow water skiers through the L-shaped structure. Over a thirteen-year period, through the height of the Great Depression, Alexander Ott's Water Follies drew crowds of 3,000 people every Sunday. Legends like Harry Houdini, Olympic skater Sonja Henie, and the Flying Wallendas were headliners at the show. Long before Olympic great Johnny Weissmuller ever went to Hollywood to play Tarzan, he was a regular at Ott's Water Follies.

Ott's son, Jackie, the Aqua Tot, was touted as the most beautiful child in the world and was winner of six baby contests, including the prestigious baby parade contests of Asbury Park and Atlantic City.

Back in New Jersey, I searched the Miami information directory, only to find that there were no Alexander Otts listed. My next option, as well as my most realistic one, was to find Jackie Ott, the former Aqua Tot. Sure enough, a Jack Ott was listed, and I wasted no time in dialing.

"Mr. Ott?" I asked.

"Yes," he replied in a monotone.

"By any chance, would you be related to an Alexander Ott?" I gingerly asked.

"I guess so," he replied. "He's my father."

Jackpot. But in that instant I questioned myself and thought, "Who says that the attack on Vansant meant enough to Alex Ott for him to share it with his family?" Or, knowing the ultimate reliability of newspaper accounts, was it really Ott who was involved with the rescue? Nevertheless, I pushed on.

"When your dad used to be a lifeguard in New Jersey," I started, "I believe he helped a man who was involved in a shark attack. Did he ever tell you that story?"

"Sure he did," Jackie replied without hesitation. "I'll tell you the story if you want me to."

He continued enthusiastically. "My dad had been patrolling the beach one late afternoon when he saw a friend of his struggling just beyond the waves. When he got out to him, he saw a big shark still biting him. The sharked remained right next to them, and the man was swimming in a pool of blood. The shark followed them all the way to shore. When they got up on to the beach, my dad saw that the man's leg was badly injured, and blood was pumping onto the sand. He did the best he could with what he didn't have, and a woman was standing nearby [possibly Vansant's sister] and he grabbed the skirt of one of those old-fashioned bathing suits and ripped a piece of cloth to make a tourniquet. But the man died. . . ."

All the flesh along the back of Vansant's left thigh was stripped from the hip to the knee, leaving the bone exposed. There was also a huge gash on the right leg. Witnesses, many of whom said that they saw the shark still stubbornly affixed to Vansant's leg in about eighteen inches of water, described it as bluish gray or black, about nine feet long, weighing approximately five hundred pounds, and possessing a distinctly large triangular dorsal fin. One witness, a sea captain, described the shark as a Spanish shark (sandtiger shark) with a circumference of about forty inches. He had not seen a shark of this type on the New Jersey coast before this time. The sea captain stated that sharks are apparently attracted by the fishing grounds offshore, where fishermen routinely throw inedible fish remains overboard. He also pointed out that, a month prior, a fisherman reportedly captured a shark and found a man's leg in its stomach. Maritime literature of the day claimed that the Spanish shark was driven from tropical waters by Spanish-American War naval bombing. The newspaper quotation attributed to the sea captain may, therefore, be nothing more than a reporter utilizing a known myth to generate witness testimony.

Local newspapermen asked Vansant's sister Louise for her interpretation of the events. Louise responded in Beach Haven's *Courier*, saying, "My brother was a good swimmer and left the crowd near shore to swim out beyond the breakers. A few minutes before, he had been playing with a big dog. The crowd, believing that he was calling for the dog, paid no heed to his cries. Everybody was horrified to see my brother splashing about in the water as though struggling with a monster under the surface. He fought desperately and as we rushed toward him we could see great quantities of blood. When Charles was taken from the water, the terrible story was revealed. For his left leg had been virtually torn off."

Dr. Eugene Vansant, along with a medical student, rushed to Charles's aid on the beach. They quickly moved him to the manager's office at the Engleside and cleared the large wooden desk to create a bed. Dr. Herbert Willis of Beach Haven (who later became mayor in the 1920s) and Dr. Joseph Neff, former director of Public Health in Philadelphia, were summoned and swiftly attempted to assist in retarding the flow of blood from the pumping wound. Dr. Willis telephoned Dr. Buchanan at Community Hospital in Toms River (thirty miles to the northwest) in an effort to have the suffering Vansant transported urgently, but Willis soon made the decision that Vansant was too weak to make the bumpy journey. Vansant had already lapsed into unconsciousness near the beach and never regained enough strength to tell the story of the attack. Charles was pronounced dead by Dr. Willis at 6:45 P.M.

Besides making Vansant's the most legible of the 1916 shark-related death certificates, Dr. Willis also meticulously recorded Vansant's period of suffering to be "1 hour and 3/4." Dr. Willis reported the immediate cause of death to be "hemorrhage from femoral artery, left side" and recorded the contributory cause of death as "bitten by shark while bathing." Dr. Vansant had later stated that he felt the cause of death to be obvious massive blood loss (progressive-hemorrhagic shock), as the "shark virtually tore Charles' left leg from his body." The body was transported to Philadelphia for burial.

The southern portions of New Jersey and Philadelphia reacted to the Vansant tragedy with immediate horror. Surprisingly, however, even though it was the first recorded shark attack fatality of the East Coast of the United States, it received minimal coverage by the northern New Jersey and New York newspapers. The *New York Times* buried the account of the Vansant attack on page 18, referring to the culprit simply as a "fish" on three occasions in the short piece and reluctantly mentioned that it was "presumably a shark." Longtime Jersey Shore residents were astounded by what had transpired in Beach Haven, but since the shore's economy relied on the daily trainloads of New York visitors, no great unrest was apparent.

James M. Meehan, the Fish Commissioner of Pennsylvania, and the former director of Philadelphia's Aquarium in Fairmont, in an effort to calm the beachgoers of Beach Haven, contributed to a *Philadelphia Public Ledger* article whose headline read: BATHERS NEED HAVE NO FEAR OF SHARKS: FISH EXPERT DECLARES ONE THAT KILLED SWIMMER MAY HAVE SOUGHT TO ATTACK DOG." In the *Public Ledger* article, Meehan stated:

Despite the death of Charles Vansant and the report that two sharks having been caught in that vicinity recently, I do not believe there is any reason why people should hesitate to go in swimming at the beaches for fear of man-eaters. The information in regard to the sharks is indefinite and I hardly believe that Vansant was attacked by a man-eater. Vansant was in the surf playing with a dog and it may be that a small shark had drifted in at high water, and was marooned by the tide. Being unable to move quickly and without food, he had come in to attack the dog and snapped at the man in passing.

Meehan conceded that the shark involved in the attack could have been a genuine man-eater (great white) but thought it more likely to be a blue shark. He also specified that if the shark of this length (approximately nine feet) were a true man-eater, it would certainly have taken Vansant's leg completely off. Meehan also mentioned that if the shark were a hammerhead, it would have easily been identified by the observers. In closing, he stressed that all sharks, dangerous or otherwise, belong in deep water and rarely come in close to shore.

These theories were corroborated by those considered to be shark experts. Some spoke of the unclaimed $500 Oelrichs reward and spoke with firm skepticism about the prospect of a true unprovoked shark attack. Others, like the scholars at the American Museum, thought the event so unique, bizarre, distant, and questionable, that "no comment" was the best comment. Many experts believed that the question of analysis would disappear as the scientifically volatile event became a blurry memory.

In preparation for the Independence Day crowds, Robert Engle, owner of the Engleside, announced that Beach Haven would protect its bathers by placing wire netting in the surf, three hundred feet distant from the shoreline and extending the entire length of the beach. Local cottage residents generated a fund to help install the nets and the local pound fishermen (pounds were immense netted containers situated four hundred yards offshore, in which fishermen would capture and store live catches) placed the pilings in and hung steel nets.

The nets were similar to those used on naval torpedo destroyers. The hastily constructed nets would last until late August when the first fierce northeasterly storm blew in.

Back in West Philadelphia, preparations were made for Vansant's funeral, scheduled for Wednesday, July 5. The mass was to be conducted from Vansant's home on Spruce Street at 11:00 A.M. under the guidance of Charles's mother, Louisa, and his father. The funeral services were open only to friends, relatives, and members of the Class of 1914 of the University of Pennsylvania. The Reverend Archibald McCallum from Vansant's Walnut Presbyterian Church was to officiate. The interment took place at the picturesque South Laurel Hill Cemetery, just outside Philadelphia.

Charles Vansant was gone, as was the male legacy of the Stophel Garrettson Vansants of Holland. The untouchable mystique surrounding the Oelrichs reward was gone as well. Today, a visit to Vansant's grave reveals that he was laid to rest at a family plot. I'll never forget the phrase that the helpful cemetery caretaker used as he pointed to a map to show me Vansant's location: "You'll find your loved one right in this section." Vansant had predeceased other family members by several years.

DIES AFTER ATTACK BY FISH

C. E. Vansant Had Been Bitten While Swimming at Beach Haven.

Special to The New York Times.

BEACH HAVEN, N. J., July 2.— Charles Epting Vansant of Philadelphia, who was badly bitten in the surf here on Saturday afternoon by a fish, presumably a shark, died late last night.

He was less than fifty feet from the beach and was swimming in when those on the shore saw the fin of a fish coming rapidly toward him. They called to him to hurry and yelled warning at him but before he swam many feet the fish closed with him. Vansant shouted for help and then went under. Alexander Ott, an expert swimmer and a member of the American Olympic swimming team, dashed to his assistance, but arrived too late to prevent his being bitten. After a struggle, Ott brought him ashore.

Mr. Vansant was the son of Mr. and Mrs. Eugene L. Vansant of 4,038 Spruce Street, Philadelphia, and was in his twenty-fifth year. He was a graduate of the Department of Fine Arts of the University of Pennsylvania and was connected with the firm of Nathan Folwell & Co., of Philadelphia.

It is said that large sharks have been seen recently a few miles out.

The New York Times, *July 2, 1916.*

CHAPTER **2**

Spring Lake: July 6

As Charles Vansant's family was preparing for funeral services, the Jersey Shore was preparing for an unforgettable Independence Day. Thousands of New York, North Jersey, and Philadelphia visitors arrived by the trainload at destinations including Wildwood, Belmar, Ocean Grove, Point Pleasant, Long Branch, and Asbury Park. While the large, commercial resorts did their best to accommodate beachgoers, exclusive getaways like Spring Lake, Allenhurst, and Deal welcomed the Atlantic coast's high society.

Back in New York City, however, the rate of mortality from infantile paralysis was running at one death an hour, and 756 cases were known. North Jersey towns were now experiencing their first cases of polio, and New York State was already demanding that all children be banned from all public gatherings. While the sweltering July heat caused fatalities in cities still devoid of air-conditioning, the reliable southeasterly ocean breeze provided Jersey Shore beaches, boardwalks, and verandas with a refreshing salt-cooled air. Little was spoken about the tragic event in Beach Haven three days before. Gaiety, splendor, and excitement prevailed. However, subtle alterations in arrangements and agendas told the astute observer that something wasn't quite right. The ocean-side pools, like Asbury Park's immense natatorium, were unusually crowded, even though the surf temperature was comfortably refreshing. Lifeguards now spent more time on their Hankins rescue boat patrols just outside the lifelines, and they became especially adamant in enforcing restrictions on the daring swimmer who might venture far from shore. While the lifeguards kept a quiet vigil for that triangular black fin, the beach managers of the sophisticated resort towns were unobtrusively inquiring about the purchase of steel submarine snare netting to place in the surf, just in case.

In 1916, no East Coast town was more elegant or opulent than that of Spring Lake, New Jersey. Situated forty-five miles north of Beach Haven, Spring Lake was a hamlet of thirty-nine hundred people, magnificently adorned by classic Victorian homes and eclectic hotels. Today, its relatively well preserved architectural state and character have placed Spring Lake's property values among the top fifteen in average price in the entire United States. In 1916, Spring Lake was one of the most visited summer playgrounds by the nation's ultra elite. Along its beachfront, two auspicious hotels, the New Monmouth and the Essex & Sussex, stood adjacent to Ocean Avenue and could easily compete with the grandest of the country's hotels. The New Monmouth Hotel was New Jersey's version of Florida's Breakers, while the Essex & Sussex, with its fantastic central domed tower/balcony, stood as a landmark for all oceangoing spectators. The Warren Hotel, on Warren Avenue of course, was arrayed with not one tower, but one on each end of its spectacular rectangular shape, which stretched half the block.

On July 4, 1916, the grand ballroom of the Essex & Sussex (E & S) welcomed some of the most influential state and national figures including secretary of the Treasury, William McAdoo, who had just arrived and settled in at his summer home on the southwest corner of Passaic and First Avenue, only two blocks from the E & S. President Wilson's personal secretary, Joseph P. Tumulty, whose 1916 summer residence was on Sixth and Park Avenue in Asbury Park, was also present and brought the news that President Wilson was to make either Long Branch or Asbury Park the location of his summer White House. New Jersey's governor, James Fielder, made his own appearance at the Independence Day celebration, and as the majestic white double doors were opened for his entrance, tremendous blasts from the Asbury Park and Belmar fireworks could be heard thundering outside like bass drums. The governor was joined by some of his more prominent cabinet members. Among the governor's staff in attendance was Dr. William Schauffler, surgeon general for the New Jersey National Guard. Schauffler did not arrange for a summer cottage in the area but chose to remain as a guest of the E & S.

Also in attendance was Mrs. George W. Childs. Childs was the widow of wealthy nineteenth-century publisher of the *Philadelphia Public Ledger*, George W. Childs. Mrs. Childs was the classic socialite and had an outgoing personality that would rival *Titanic*'s Molly Brown. Her late husband was not only an owner of much shore property, but he was also responsible for inducing President Ulysses S. Grant to make Long Branch (eight miles to the north of Spring Lake) his summer home and the premier summer resort in America. In a thank you letter to Childs, President Grant said, "In all of my travels I have never seen a place which was better suited for a summer residence."

During those years, it was also not unusual to find a polished German naval officer at a hotel affair. Since America's position on the world war was neutral, visits from German U-boats into Northeast ports was not uncommon. Ironically, it was the shipping company of the late Hermann Oelrichs, the North German Lloyd Line, which made the U-boats' journey possible. The German soldiers were greeted at social affairs with a combination of curiosity, courtesy, and suspicion. Many of the visiting German military men were quoted as saying that they viewed Americans as simplistic. Their presence would often prompt many to comment on how lucky the United States was not to be involved in the ugly battles of World War I raging overseas.

Two Essex & Sussex staff members also present were the stocky, twenty-eight-year-old Swiss bell captain, Charles Bruder, and the elevator runner, Henry Nolan. Both men would have been thoroughly visible for any guest who needed seating assistance. Nolan was well liked, but Bruder's personality made him a Spring Lake favorite. He was an excellent bellhop and had worked for the Spring Lake hotels since the age of eight. The winter before, Bruder's hotel employment even took him to the Hotel Green in Pasadena, California.

Everyone, especially the shore merchants and local political figures, was relieved and optimistic that the summer would turn out to be one of the most splendid in memory. Even so, the sea captains who entered the ports of Newark and New York continued to tell tales of large schools of sharks just off the New Jersey coast. Those who were generally optimistic thought little of the warnings. During this summer, however, they should have realized that all bets were off. In fact, what would transpire on the afternoon of July 6 would send shock waves through the overly calm and jubilant community of Spring Lake.

Staff, including Bruder and Nolan, who were forced to work through the oppressive daytime heat, would resort to shade, perhaps in a cool basement, or would try to take advantage of the warm puffs of air that blew, like clockwork, every day between 1:00 and 2:00 P.M., transforming stillness into a refreshing flow of ocean-cooled breeze. For most of the week, the daytime temperatures hovered in the low eighties. The wind had been out of the north during the early afternoon, and the ocean water was relatively warm for early summer. Toward the end of the week the wind began to howl from the south even earlier than the 1:00 to 2:00 o'clock transformation. The southerly wind in early July was a recipe for cold ocean water. On July 6, Charles Bruder and Henry Nolan planned their lunch break at 1:45 P.M., as they did almost every day. After sweat-filled mornings of lugging guest bags, Bruder and Nolan used their lunch hour to cool off in the sea.

On their short jaunt over the boardwalk toward the employee bathhouses at the South End pavilion, Bruder and Nolan noticed that it was midtide (outgoing) and that the guest's beach was unusually crowded, with an exceptional array of elegant bathing beauties. Some of the ladies were wearing plain, black, baggy, long dresses with leggings; others, the boardwalk-bound individuals, were wearing frilled white dresses with beige bonnets. Although the men gazed at and commented on the attractiveness of some of the ladies, they kept their pace lively; they, too, wanted to take advantage of the inviting surf. Before they made it to the bathhouses they were confronted by a wet-headed young man named Robert Dowling, who was just returning from an hour-long swim that had taken him a good distance from shore. The men asked Dowling how the water was. "At first it's a bit chilly," Dowling responded, "but after that you get used to it or get numb."

Not only was Charles Bruder considered the most likable employee by hotel guests, he was also known as a fearless swimmer. Bruder scoffed at the talk of sharks after the Vansant death by citing his experience with timid large sharks off the California coast only months before. He enjoyed his afternoon swim more than anything else in his day and was always the first in the water and the one to outdistance other bathers. His daily lunch break was well deserved; Bruder was a devoted son who sent much of his income directly to his mom in Lucerne, Switzerland. Charles's brother, back in the old country, had been fighting the battles of World War I, and it was up to Charles to support his mother during this period.

Bruder and Nolan reached the south end pavilion bathhouses, and Nolan took one last glance toward the bathing area just east of the hotel. Some ladies were gingerly making their way into waist-deep surf, and like most of the bathers, they held on to the swim ropes perpendicular to shore. Many bathers would crouch playfully—"fanny dunking"—with each oncoming wave and giggle with ecstasy as the foamy surge and salt spray covered their bodies.

Some of the male bathers were wearing striped or plain white tops, and some had cloth belts that were sewn right into the fabric. Henry and Charles were carrying their suits, which were the standard all black tank-top type. All bathers were required to utilize the bath-houses; it was a major violation of etiquette to walk on the boardwalk while wearing bathing attire. Just before the men entered their respective changing lockers, Nolan joked with Bruder, saying, "What if I dove under and came up in front of one of those pretty young ladies with my black swim tights on?! They'd think I'm a sea monster, or at least the real Moby Dick," he boasted. Charles was less than amused, and responded, "If you want to keep your job, you better stay clear of the guest's bathing area." Just as they were entering the lockers, Henry spotted Childs peering off her private balcony with theater glasses. "You're right, Charles," Henry said with relief. "Mrs. Childs would have had my head for sure if I swam anywhere outside the employee section."

On duty at the south end pavilion beach were veteran lifeguards Chris Anderson and Captain George White. George White was so prominent in the town's services that he doubled as beachfront manager. Bruder and Nolan joined some of their friends in the surf at about 2:15 P.M. As usual, Bruder's friends became chilled and tired much sooner than he did. The air was warm, but the south breeze was cooling the water to the low sixties. Those who know the New Jersey shoreline currents recognize that wind and water temperature are closely linked. The net current from Bay Head south is toward the south, and from Bay Head (which lies several miles below Spring Lake) north, the current flows north. The currents from the south bring notoriously frigid water, while the currents that arise with a northeast wind (from the Northeast) bring incredibly warm seas. The north or northeast winds seem to provide drifts from the warm Gulf Stream to flow toward the beach.

When Charles's friends left the swim area, he used their departure and the fact that the inshore water was on the low side (it was only two and a half hours until low tide) as an excuse to venture far beyond the lifelines, approximately 130 yards from shore.

Up on shore, beachgoers were toweling off or dozing under black-and-white-striped umbrellas. In the E & S hotel dining room, Dr. William Schauffler was just finishing lunch with the hotel doctor, Spring Lake's renowned local physician, Dr. William Trout. The afternoon tranquility, however, was suddenly interrupted.

Without warning, two strange shrieks followed by a deep scream were heard from the water. A woman standing near the lifeguard stand suddenly pointed in the direction of Charles Bruder and shouted to the lifeguards that she believed that a canoe had capsized just outside the swim area. The woman emphasized that the canoe's hull was just at the surface of the water and painted a deep red. This, she asserted, could be the only explanation for the water's blue hue to change radically to a crimson red. White and Anderson did not recall seeing anyone on the water with a canoe. Instantly, they realized that something was very wrong. What they were seeing was not the red hull of a canoe but a swirling circle of blood-red water, and Charles Bruder was struggling and splashing amid it.

Since the death of Charles Vansant, White and Anderson had thought about the possibility of a shark attack, but there was no real way for them to be prepared. Except for Alexander Ott five days earlier, no U.S. lifeguard had ever responded to such a form of distress. The guards quickly launched the rescue boat and rowed with the intensity that only adrenaline and instinct can generate. They took intermittent looks toward Bruder and saw his head bobbing just above the surface. Their objective was clear; get to Bruder before he went under.

Scores of people began gathering on the beach, some having come from several blocks away. One witness claimed that Bruder's bloodcurdling scream was heard from a quarter mile away. Another mentioned that Bruder was flung high in the air between strikes. Henry Nolan and Bruder's other friends now waited anxiously along the shoreline. Mona Childs, who had seen the entire event unfold through the lenses of her theater spectacles, rushed down to the hotel switchboard and demanded that the operator contact her niece in Deal, as well as the other hotels up and down the coast. The message: "Get out of the water!" Childs would later explain that she'd seen the frightful shark dart toward poor Bruder, turn away, then dart back at him "just as an airplane attacks a Zeppelin." Another eyewitness account reported that Bruder "leaped from the water with his right leg gone from above the knee, and blood spurting from the wound. He then fell back into the water and the shark made another attack, this time severing his left foot." Regardless of the sensationalized gruesome legless acrobatics that Bruder was said to have performed, the multiple strikes at Bruder, we will soon find, certainly did occur.

When White and Anderson were just yards from Bruder, they could see the ominous pallor to his face and the desperate effort he was making to stay afloat. With his life slipping toward heaven, Bruder shouted, "A shark bit me! Bit my legs off!" White extended an oar to Bruder, and Bruder unsuccessfully attempted to pull himself into the boat. White then grabbed Bruder's outstretched arm and forcefully hoisted him over the gunnel. White was surprised to find that Bruder was much lighter than he had anticipated. The men could not conceal their horror as it was revealed that Bruder's lower legs were torn off just below the knees. During the brief return trip to shore, Bruder, lying on the bottom of the lifeboat, gazed up at the blue sky as his little remaining blood filled the deck. He lost consciousness almost immediately, and the lifeguards began rowing a tad slower. It was obvious that life had departed from his mutilated body.

On the beach, Dr. John Cornell, the house physician at the New Monmouth Hotel, and Dr. William Trout were summoned by the E & S manager. When the doctors arrived at the scene, the hotel manager requested that they defer examining the mutilated corpse in order to care for the scores of women who had fainted and vomited at the sight of Bruder's remains.

The nature of Charles Bruder's wounds, which clearly included the severing of multiple lower extremity arteries, produced such catastrophic blood loss that he died of irreversible circulatory (hemorrhagic) shock within minutes of being hauled into the lifeguard boat. While doctors Cornell and Trout were reviving women who had swooned, Dr. Schauffler, of the governor's staff, examined Bruder's injuries fifteen minutes after death. Schauffler found that much of the flesh on Bruder's right leg was torn off, and the lower leg bones (tibia and fibula) were severed halfway between the knee and ankle, with the remaining muscles hanging in shreds. The left foot was missing, and the left lower leg bones were also severed but at a lower point than on the right. The bones that remained intact on the left leg were largely denuded of flesh and muscle, and those that remained were also torn and jagged. Dr. Schauffler also noted that the skin, muscles, fat, and bones that were affected revealed shredded ends. A deep circular gash down to the thighbone (femur) above the left knee was also apparent, as was a gouge, which measured about "the size of an apple or a fist," reaching down through skin, muscle, and fat at the right side of the abdomen. The peritoneum (abdominal lining covering the intestines and other organs) was not opened.

On the beach was Robert Dowling, the nineteen-year-old son of the president of a City Investment Company in New York, and the bather who had talked to Nolan and Bruder only minutes before their swim. He was an accomplished long-distance swimmer and had done the forty-mile swim around Manhattan the year before. Only twenty minutes before the Bruder attack, Dowling had returned from a two-mile swim straight offshore. His return path had taken him directly over the area in which Bruder had been attacked. He must have been relieved that he was still alive.

The media response to the Bruder attack was one of absolute shock and amazement. The Vansant attack of July 1 had seemed like a quiet, if freakish, aberration, but when the unthinkable attack on Bruder occurred in front of hundreds of tourists, including the upper crust of society, the cork burst off the magnum bottle of champagne. This was the blast after the salvo.

The Spring Lake shark attack was suddenly stealing headlines from the war. The *Boston Herald,* the *Chicago Sun Times,* the *Washington Post,* and the *San Francisco Chronicle* all ran the front-page stories. The *New York Times* also ran the article on the front page, and this time they made no attempts to waffle on the identity of the attacker. The *Times* headline read: BATHER HAS BOTH LEGS BITTEN OFF BY SHARK ON THE JERSEY COAST: CHARLES BRUDER ATTACKED WHILE BATHING AT SPRING LAKE YESTERDAY AFTERNOON IN SIGHT OF 500 PEOPLE. The scientists at the American Museum of Natural History were now unable to ignore the reality of the shark attacks as they were thrown directly into the media frenzy. Despite the clear but questionable position of these scientists on the probability of unprovoked shark attacks, the U.S. and New Jersey governments, as well as the general media, still looked to these men as the experts on the topic.

On July 8, a press conference convened at the American Museum in Manhattan. Present for questions were Dr. Frederic Lucas, sixty-four, director of the museum, Dr. John T. Nichols, thirty-three, curator of the Fishes Wing, and Robert Murphy, twenty-nine, director of the Brooklyn Museum. As was mention in the Introduction, in April of 1916 (only three months prior), these three men collaborated on a scientific journal article in the *Brooklyn Museum Science Bulletin* which described the natural history and behavior of the sharks of Long Island waters. The journal article spoke at length of the great improbability of a true unprovoked shark attack against a live man, and Lucas even went so far as to say that a shark was incapable of severing human bones. In the piece, Murphy noted how hard it was to cut a leg of lamb with a serrated steak knife.

Even though Dr. Lucas's April assertions about sharks would now seem to make him an unlikely choice to head a shark attack investigation, his reputation as a naturalist and museum director was impeccable. In 1904, Lucas singlehandedly transformed the Brooklyn Museum into an organized and impressive hall of exhibits. His idea about natural history museums was that they should be interesting educational centers, not merely repositories for curiosities. Professor Henry

Osborn, president of the American Museum, once wrote of Lucas that he was "a man of fine personal character, of quaint and persuasive charm, representative of the finest principles and endowments of the New England character." He was also known as a strong but kind leader who would win controversies without losing friends. The staff at the American Museum, where he took over as director in 1911, knew him as a very modest, very shy individual, but firm as a rock.

When the American Museum's conference room door opened, one reporter immediately asked Dr. Lucas what he would do for the thirty Jersey Shore mayors who were asking the scientists to thwart fears of the panicked vacationers. Lucas was revered as the top-notch scientist, which he was, but if the situation got any more grave or controversial, John Nichols would most likely have to run the show. After all, Lucas was an anatomist, Murphy was an avian zoologist, and Nichols was the true ichthyologist. Murphy and Lucas jointly stressed the great unlikelihood of a repeat attack, and when they were asked if they could ensure the future safety of bathers on the Jersey Shore, Nichols spoke of the third-of-an-inch steel-wire-mesh fence up at Asbury Park's bathing facilities and many other resort bathing areas. Nichols felt that the netting and caution about venturing too far from shore should stem any further danger. The experts took pains to point out that they were as surprised, if not more so, than the general public, especially in view of the fact that the Oelrichs reward went unclaimed for twenty-five years. When Lucas was asked about the possibility of a beast other than a shark being the cause of the New Jersey fatalities, other reporters chimed in with suggestions of killer mackerel or monstrous sea turtles or even German U-boat torpedo being pinned as the cause. "I'd prefer not to jump to any conclusions," he responded. Lucas was aware that the Coast Guard and the Marine Fisheries Service were waiting for an eyewitness description of Bruder's wounds from Dr. Schauffler, the surgeon general for the New Jersey National Guard, before they made any comment. Lucas was promised a copy of Schauffler's report and stated that he would withhold final comment on the exact perpetrators until he could better evaluate the evidence.

Despite his earlier statement that any shark to venture into shallow waters would be the nonaggressive sort, James Meehan, former Commissioner of Fisheries in Philadelphia, did not hesitate to incriminate a shark as the guilty party. Meehan, now forcefully convinced of what he had doubted earlier, humbly asserted that a single monster shark that drifted up from the tropics was responsible for the attacks. Likewise, in Dr. Schauffler's assessment of the Bruder wounds sent to Dr. Lucas, Schauffler wrote: "There is not the slightest doubt that a man-eating shark inflicted the injuries." Dr. Schauffler, who later became a major figure in New Jersey's Medical Society, became an aggressive proponent of wire mesh at all bathing areas at shore municipalities. Despite Dr. Schauffler's words of urging, seasoned seamen and many other "experts" were quoted as saying that they felt that the Bruder incident was not only a once-in-a-lifetime freak occurrence, but that such a tragedy could not happen again in a *thousand years*. A thousand years would be thirteen or fourteen lifetimes. The Spring Lake doctors, Trout and Cornell, who examined Bruder after his corpse was taken to the New Monmouth Hotel, reported a detailed assessment of his wounds to the district superintendent of the U.S. Coast Guard, John Cole. "On examination, we found that both legs were missing; bitten or broken off about 4 inches above the ankles and a large cut above the knee. The flesh (was) torn off the right leg from just below the knee to where the bone was bitten or broken off, leaving the bones protruding without any flesh. A piece of flesh bitten out of right side below the ribs, also showing tooth marks. We are of the opinion," they concluded, "that these injuries were caused by a shark. In our judgement we do not think there is any doubt." Coast Guard superintendent Cole, however, was no great help in promoting appropriate caution among bathers. In a *Philadelphia Ledger* article, he described sharks as timid as "rabbits." Cole also mentioned that he frequently used to swim through schools of them in Cape May. "If they got too thick or close, you would just throw a clam at them." In addition, U.S. Fisheries Commissioner Hugh M. Smith presented his optimism about safe bathing by insisting that there would be no further attacks. Smith felt that the shark or sharks involved must have been unusually hungry, and noted that there are thousands of people who bathe each year yet not one single attack was recorded before the summer of 1916. It was a record in favor of sharks, "a record of safety that domestic animals, like dogs, cats, and horses, could not even match."* Dr. Smith added that the scarcity of men-haden might be the cause of unusual hunger in sharks just like "turn-ing a dog mad or a mule kicking." Finally, Smith stated that he was not altogether convinced that a shark was responsible for either of the fa-talities. He felt there was a remote possibility that swordfish might have strayed into the bathing areas.

Regardless of the fact that academic authority considered the possi-bility of another shark attack to be extremely remote, shore resorts were not taking chances on the mind-set of vacationers. After the at-tack on Charles Bruder, it took only twelve minutes to clear the water from Spring Lake to Cape May. At places like Asbury Park, where rail-way cars were routinely unloading thousands of New York vacationers every few minutes, the Chamber of Commerce had to do some quick thinking to salvage summer tourism. The *Asbury Park Press* headline of July 7 attempted to help the situation with its assertion of swift and ef-fective action: NETS AND ARMED MOTORBOATS TO PROTECT BATHERS: BE-LIEVED PRECAUTION TAKEN WILL ASSURE ABSOLUTE SAFETY TO BATHERS ALONG NORTH SHORE. ASBURY PARK BATHING GROUNDS ALL TO BE SUR-ROUNDED BY WIRE—THINK REPETITION MOST UNLIKELY. Six weeks of pre-cious summer remained, and resort owners were optimistic about making up for a few days of lost business.

Russell Cable was a nineteen-year-old lifeguard at Asbury Park's Fourth Avenue beach in 1916, and in another rare interview opportu-nity, Mr. Cable spoke of excursion trains, ten to twelve cars long, pulling into the North Asbury station every minute and half. "As soon as one would get four blocks away," Cable recalled, "there'd be another one pulling in, unloading thousands of people. I've seen lines three across and a block long standing out front of the beach ticket shack

---

*Asbury Park Press, July 1916.

waiting to pay the twenty-five-cent admission." Cable didn't seem to think much of the effectiveness of the wire netting but stated that he believed it was installed to keep people within the swim poles, and if a lifeguard spotted a fin cutting the surface of the water, "he wouldn't have far to go to achieve a rescue." According to Mr. Cable, the wire fencing was put up to "soothe the feelings and the thinking of the people of New York. They had a safe place to go swimming." One should remember that, although Asbury Park today is in need of a major structural renaissance (extensive waterfront redevelopment is planned for the fall of 2001), the Asbury Park of 1916 and the towns just to its north were the places to be for anyone who was anyone. In other words, it was one thing for Charles Vansant to reach his demise in the less traveled southern portion of the state, but when a legless victim was snatched in the North Shore, the wealthy New Yorkers demanded action.

At Asbury Park and many other locations, motorboat patrols were established to locate sharks offshore. Each boat was equipped with rifles, axes, harpoon guns, and long lines baited with huge pieces of lamb or quarters of sheep. The steamers coming in from overseas were even catching sharks using steak as bait. Oliver H. Brown, mayor of Spring Lake, established three shoreline patrol boats to cruise the beachfront. On July 8, the armed boats went on a wild mission when someone spotted a dorsal fin near the bathing area. Shots were fired and the fin chased. After the commotion, it still wasn't clear if they were shooting at a porpoise or a shark.

At Asbury Park's Asbury Avenue swimming area, a shark was sited just outside the swim ropes. The lifeguard captain there, Benjamin Everingham, was forced to beat with an oar on the head of the reportedly twelve-foot shark. The shark fled to open sea after several blows, but after the encounter, all bathers were warned to swim only at the enclosed bathing area at Fourth Avenue. The entrance gates at the unfenced bathing grounds were all locked. Everingham's "shark-oar story" not only made the local papers, it made national front-page news.

Soon, the Atlantic coast was alive with battles against man-eaters. Columnists and writers from all types of publications descended on the seemingly shark-infested Jersey Shore. A columnist for *Field and Stream* was portrayed capturing a large sandbar shark in the surf just off Beach Haven. A camera was at the ready, of course, when the shark was landed by E. F. Warner and his assistant, New York sportsman Herbert Savage. The men got the shark to the beach in thirty-five minutes and then put three .38-caliber revolver shots in its head as it splashed in waist-deep water. The shark was a dusky or a sandbar shark, and it measured nearly six feet long and weighed ninety pounds (twelve hours after it was caught).

The shark hunt was not confined to the shore region alone. A policeman in Bayonne, New Jersey, emptied his revolver into a huge black fin lurking near a yacht club dock, and the magnificent shark was depicted in the newspaper with its captor. The photo of the Bayonne shark is more like an illustration than a photograph, and, to me, it appears to be a cross between a great white and a thresher shark. In another interesting capture, fishermen digging for sandworms off Rocky Point, New York, spotted a large shark chasing weakfish toward shore. They quickly utilized eel tongs, oars, spears, and spades to slash the shark to death as it approached a sandbar.

While the attacks on Vansant and Bruder were anything but amusing to shore inhabitants, the newspaper cartoonists around the country were having a field day utilizing a newfound foil for caricature. Some cartoonists used the shark as a symbol of unsavory political figures or unpopular institutions. Since 1916 was among the years that Americans were trying to break away from the rigidity and conservatism of the Victorian period, one comic depicted a risque polka-dot bathing suit and advertised it as the secret weapon to keep sharks away from our swimmers. Another cartoon portrayed the reality of the summer's nightmare quite succinctly by depicting an exasperated individual at the end of a dock that displays a DANGER: NO SWIMMING sign and mentions the three most emphasized "danger" topics of the day: "Infantile Paralysis (polio), Epidemic Heat Wave, and Sharks in the Ocean." The cartoon was titled "What's a family man to do?" In those days, the New York papers were also inundated with advertisements for New Jersey Shore resorts. One such advertisement read: "Come down and laugh at the sharks—we have enclosed our bathing area with reinforced steel nets!"

More amusing and lighter discussion ensued in Spring Lake when a druggist opened a letter from a man from Hot Springs, Arkansas, requesting shark oil. The letter had gone to the local post office addressed to "Any Reliable Pharmacy in Spring Lake." The man from Arkansas, apparently believing that Spring Lake was being overrun by sharks, requested that a "pure grade of the shark oil be sent." He even provided payment prices by the ounce. The oil was said to have intense medicinal properties.

The idea that there was a plentitude of sharks off New Jersey beaches, however, may not have been simply pure imagination or the product of resourceful journalism, according to Dr. Robert Patterson (age ninety-five during an interview in 1989). A lifelong resident of Spring Lake, he recalled the wire-and-chain meshing that was placed at the bathing grounds at the local beaches. "A funny part of it was, I remember so well," Patterson said, his eyes bright and vibrant as he spoke, "seeing a little shark caught right up in the chain. We knew the sharks were out there, and we knew people were getting killed by the sharks. I recall that they took an old harpoon down from the ceiling at one of the beach pavilions and they went out off the beach in a boat and harpooned this shark. My God, it took them for a buggy ride! That shark took them around and around and around, you know. Oh heavens, the water was full of them."

As a Jersey Shore historian and resident, I relish any opportunity to get my hands on old photographs or antique postcards, especially of the central or northern coast. Postcards of the turn of the century are much different from those we have today. Since color photography was not yet available at that time, many of the fine postcards of the period, most of which were actually produced in Germany, were initially black and white photos that were hand dyed with colored paint. The result was a spectacular and idealized image. In 1989, I attended a large postcard show in Belmar, New Jersey. Rarely do I ever read the back of a card (I'd never make it successfully through a show were I to spend more than a few seconds examining every single card). However, to better date the card I do flip it over to check out the postmark. On this day, I turned a card that showed a standard Asbury Park bathing scene. I was looking for the style of beachwear that was consistent with the 1916 period. The card was postmarked "JULY 8, 1916." I couldn't resist reading the note. The card was sent by a honeymooner named Mona to a friend in Ludlow, Massachusetts. Sure enough, the writer informs Mrs. O. D. Tucker that their bathing beach has been screened in and that the water is being patroled by boats "since the scare of a shark biting off the legs of a man a few beaches above here. . . . The man died. Since then, a great many bathers are rather scarce." The sender of the Asbury card also mentions that, since she was married (ten days prior), there had not been one unpleasant day of weather. Another rare postcard was recently recovered by the director of the Asbury Park Public Library, Robert Stewart. The card was sent from Asbury Park by a woman named "Peggy" in mid-August, 1916, to a friend in Ohio. Peggy wrote that, as a woman, she was safe from the New Jersey sharks because they were "the *man*-eating variety."

Back in Spring Lake, Mrs. Childs suggested that a fund be taken up to ensure that the beloved Charles Bruder had a proper burial. Bruder was so admired by the Essex & Sussex hotel guests that they donated the money needed for his funeral. The hotel guests at the Monmouth Hotel, the Breakers, and the cottages also contributed to the fund. The E & S hotel management insisted on bearing the total expense of the fund and planned to turn it over to Bruder's mother in Lucerne, Switzerland. Dr. Charles Vuilenuier, Swiss consul in Philadelphia, said that it would be better to inter the body in the United States since the war would make shipment difficult. He also arranged for the money to be sent through the Swiss embassy in Washington. Charles Bruder's coworkers contributed a heartfelt $1,000 to send to Charles's mother as well.

Since I was never able to track down Bruder's death certificate, I went on a mission to find something more substantial than newspaper accounts to verify his death. Unfortunately, Bruder's medical examiner records were either never generated or have long been discarded. To make matters worse, sometimes dates cited are inaccurate. Charles Vansant's death, for example, is commonly reported as July 2 as opposed to July 1. Presumably, the date on which the story appeared in the newspapers (the day after the attack) was the day authors/researchers reported for the incident. Even the International Shark Attack File report is inconclusive as to the correct spelling of Bruder's last name. They record it as Bruder or Pruder. These are relatively minor facts, but they can be of major significance during research requests. Fortunately, in reference to Bruder, I knew that after a funeral mass at St. Andrew's First Methodist Church in Manasquan, he was interred at Atlantic View Cemetery in Manasquan (just south of Spring Lake). After a whole day of walking the cemetery without any luck, I gave up and contacted the cemetery's administrative office with a request. Within a day, my message machine at home had recorded a short message from the Atlantic View groundskeeper: "I found Bruder for you." The kind gentleman also gave simple directions for locating the plot; the following day, under a shady tree, I found Bruder's modest tombstone. It was rectangular gray marble, bearing the simple inscription, CHARLES BRUDER JULY 6, 1916. The lone stone does attest to Bruder's independent existence in the United States and, more important, it confirms the spelling of the victim's name and the date of his death.

In the late eighties, when I was doing research for the first comprehensive video on the subject, *Tracking the Jersey Man-Eater*, I contacted Alva Allen, a lifelong resident of Spring Lake who was now living with her nephew. At ninety-five years of age, she was none other than the grammar school classmate of Dr. Robert Patterson. The woman was spry enough to correct me on a few minor points during the course of the telephone conversation, and she immediately remembered Bruder's tragic death "as if it were yesterday." "Oh yes," she recalled, "he was laid out right next to the Five and Dime that I was working at on Main Street. It was a closed coffin, you know." I spoke with the pleasant woman for several minutes and then arranged a video interview for the following day. When I got to her home at Spring Lake, her gray-haired nephew answered the door. I explained the interview arrangement and the telephone conversation of the night before, as I struggled to keep my equipment balanced. The nephew looked at me with a blank expression and said, "Well, that's all you're going to get." The night before, he explained, his aunt had just finished telling him of the shark attack story (prompted by my call) and then dozed off. When he checked on her some hours later, she had passed away.

Just days after Bruder's death in Spring Lake, the E & S hotel had a large reception to honor Treasury Secretary McAdoo. While President Wilson, a former New Jersey governor and state resident, was finalizing his decision about the location of the summer capital, he was also in direct communication with McAdoo and Joseph Tumulty. The president wanted to appear on top of things when it came to appropriate and decisive action on the shark attacks. Wilson realized he could suffer political disaster, especially with the election only four months away, if he did not show sincere concern for his home state from the affliction of economic disaster at the hands of a man-eating shark.

By July 10, the smoke was finally beginning to clear on the news of the attacks. Most of the United States was preoccupied by the activities of Poncho Villa and his bandits. By this time the outlaw Villa had destroyed government forces in Mexico at Corralitos and Chihuahua, and Texas had already been violated by his growing band of raiders. The *Washington Post* reported on its front page that bathing at New Jersey resorts had resumed to a reduced rate of a "low ebb," and the northern New Jersey newspapers reported: ASBURY NOW FEELS SAFE FROM SHARKS. The subtitle of the article mentioned: NET PROTECTORS RECEIVE FINAL TOUCHES AND STEEL BARRIERS ARE STARTED AT OCEAN GROVE.

Robert Dowling, the long-distance swimmer who ventured in from a swim minutes before Charles Bruder was attacked, announced that he still planned to perform his July 29 swim from Battery, New York, to Sandy Hook, New Jersey. Dowling's dad, however, was preparing a thirty-foot-wide floating wire "basket" to protect his son during the swim.

C. R. R. of N. J. Station. Matawan, N. J.

Matawan train station. Early 1900s.

CHAPTER **3**

Matawan: July 12

As hoped, by Tuesday, July 11, record crowds flocked to Jersey Shore beaches. New York beaches were also robust with bathers. At the Manhattan Baths, at Manhattan Beach, the proprietor put up a five-hundred-foot-long, twenty-four-foot-high steel net and concrete walls on either end. In other places, like on the deck of the battleship U.S.S. *Texas,* eight sailors made headlines and photo ops by dispatching a fifteen-foot hammerhead while on their way to port in Rhode Island. The sailors captured the monster with a marlin spike. On July 12, U.S. officials were still having trouble gaining the cooperation of the Germans for an inspection of the U-boat *Deutschland,* and the political cartoons depicted the German lack of respect as a black submarine with the mouth and features of a shark (labeled "The Shark!") tossing Uncle Sam skyward above the water like a bull would do to a rodeo clown.

Caution and nets were the rule, but talk of sharks and shark attack had whittled down to a conversation novelty. For the inland towns of New Jersey, it was just another scorching afternoon. One such town was Matawan, located thirty miles north of Spring Lake and sixteen miles inland from the sea. Matawan, named for its pre–Revolutionary Indian heritage, was a quiet town of only twelve hundred residents. The town was almost more like a midwestern hamlet than a New Jersey village. It was surrounded by hardworking farmers, but it also possessed a quaint business district that offered standard American storefronts. Matawan's Main Street was home to Mulsoff's Barber Shop, Fisher's Dry Cleaning, Cartan Department Store, and Wooley's Hardware. In those days, multidimensional and resourceful shop owners were the people most likely to help out the neighbor in need. And in Matawan, everybody was a neighbor.

Friendliness was the norm, and such a tradition came in handy when one needed assistance in an unexpected jam. For example, the town's twenty-four-year-old blond giant of a tailor, Stanley Fisher, sometimes doubled as a baseball player whenever the local kids needed a ninth man to get a pick-up game moving. Head cop in town, Chief Frank Mulsoff, was also the town barber. His was also one of the few businesses that had a telephone. Carpenter Arthur Smith found himself applying his trade at neighbors' homes much more than at his Main Street business location. Asher Wooley owned the hardware place where one could find gardening and farming equipment just next to the sticks of dynamite. Demolition of houses and buildings probably wasn't too common at that time, but clearing trees for new construction and establishing new parks was a necessity. I understand that the creation of Central Park in Manhattan required felling numerous trees and even reconfiguring topography in major ways. Fabricating small rolling hills and gullies is not an easy thing to do. It's always better if a town happens to possess the natural beauty of a bayshore community like Matawan.

Matawan was also a servant to industry as it housed a bag factory, a tile company, and a basket factory. Young and old, parent and sibling, all took advantage of the employment opportunities presented by the factories. Young, fair-haired Lester Stillwell, for example, would report to the Anderson basket factory and Saw Mill each day as an apprentice to his dad, William. Lester's dad was half Native American and also said to be somewhat more stoic than most other fathers in town. Young Lester was a diligent worker whose summer job was to do "piece work," which consisted of nailing peach baskets together.

The residents' work ethic, however, did not curtail their efforts to treat outsiders amiably. Most noticeable to visitors was the invariable recognition that Matawan was not like most other towns. Outsiders were treated like insiders, and insiders treated like family. The people of Matawan never seemed to mind showing unconditional kindness as an instinctual element of their daily existence.

The town also held a rich history and a marvelously appealing Victorian architecture. Philip Freneau, the official poet of the Revolutionary War, once called Matawan his home, as did a group of early American seamen who served during the war as colonial privateers. The colonial privateers were patriot pirates, who were essentially former fishermen and whalers. They were absolutely essential to America's chances during the revolution and sprung into action when America most needed them. That heritage gave Matawan residents a deep sense of pride, as it was universally understood that those early seafaring citizens helped our infant nation out of a huge jam at a time when the Colonial Army was facing the British Navy, the greatest navy on the planet, and there was not yet an established Continental Navy. The heroic and extraordinary tales of these men and their maneuverable small crafts is a story unto itself, but let it suffice to say that those patriot seamen confronted the mighty British naval force head on and captured many a prize ship and contributed much to our chances for independence.

Matawan's residents were also the God-fearing kind. With four picturesque churches lining the town's Main Street, the good book was not an unknown resource. The July 9 service at the First Methodist Church, led by the Reverend Leon Chamberlain, would have seen in attendance Stanley Fisher, who lent his voice every Sunday to the choir, the pretty schoolteacher Mary Anderson, daughter of the Anderson Mill proprietor (and involved in a budding romance with Stanley Fisher), young Lester Stillwell, his best friend Albert "Ally" O'Hara, and schoolmate Charles Van Brunt. That Sunday, the Reverend Chamberlain used the tragedy of the shoreline shark attacks to speak of the Book of Genesis and God's ultimate designation for man as the master of the planet and the oceans. "Man has dominion over all living creatures and we must recognize this fact in such times of confusion." The congregants were enthralled, but Lester, Allie, and Charles, no doubt, felt every ounce of humidity in the air and could think of nothing more inviting than a couple of cooling jumps off the pilings at the old Wyckoff Dock.

While Matawan was a town linked to God, it was also a town linked to other New Jersey municipalities and Manhattan. The New York & Long Branch Railroad ran through the center of town, and a finely crafted wooden dormered and hipped train station greeted the passengers coming and going. After church services, Stanley Fisher found himself at the train station grappling with some luggage that his mother and dad were taking with them on a trip to Minnesota. Stanley's father, Watson, was somewhat disappointed that his daughter, Augusta, had moved so far away after her marriage, but the trip was a pleasant excuse to escape the blistering Northeast. Although Watson Fisher's daughter had married and moved out west, most natives of the bayshore area never seemed to drift too far from home.

Matawan's link to the rest of the world was not, however, limited to the railway. The town also had an obscure connection to the Atlantic Ocean called the Matawan Creek. The creek, or "crik" to the locals, was well known and well traveled by the former privateers and continued to be a transport route for farm produce and bricks to the New York markets and warehouses. The narrow passage was a winding bayou, forty feet across at its widest point and no deeper than thirty feet. As a brackish waterway and classic northeastern tidal river, it originated near a lake at the center of Matawan and opened about one and a half miles downstream at the Keyport Harbor and Raritan Bay.

Clam diggers worked the shallows at low tide, and at high water small boats were able to negotiate the serpentine path for transportation and delivery of goods.

On the Cliffwood side (north side) of the creek, less than a half mile distant from Matawan, at what is today the location of Garden State Parkway mile marker 119, was once the New Jersey Clay Company Brickyard docks. The brickyard would take advantage of high tide to load small barges headed for New York. In the 1800s, the creek was somewhat wider than it was in 1916 (or today, for that matter), and at that time, a tug-size steamer by the name *Wyckoff* had its own wooden dock just west of Matawan's train trestle. In 1916, only a dilapidated wharf and a dozen or so pilings remained of the Wyckoff dock, but the dock area was far from abandoned. Without a fancy municipal pool or the convenience of ocean bathing, the children of Matawan made a routine of seeking cool relief at the local swimming hole at the Wyckoff dock. One or two different small gangs of boys (girls would never swim there) segregated by general age grouping would swim and play at the dock on a daily basis. In the summer, just to escape the blistering heat and humidity, the kids swam in the nude and would often jump from piling to piling in a form of tag. A ninety-degree bend upstream and just opposite the dock was the final resting place of a schooner that had come aground on the bank at the turn of the century. The old wooden vessel sunk into the mud and became a mound for flourishing reeds, cattails, eel grass, and hermit crabs. Somewhere between the opposite creek bank and the swimming hole was an isolated deep portion, with which town boys were well acquainted. The deep spot was apparently utilized for the purposes of spectacular, if not unorthodox, high dives off the pilings. Farm kids would often want to display their swimming prowess in the deep and slightly less salty (therefore less buoyant) water. Since the women's movement was decades away from gaining any significant headway, the girls in town didn't seem to protest their apparent exclusion from the swimming grounds. Young ladies like Mary Bailey, Alfreda Metz,

Marion Smith, and Mildred Van Cleaf never actually went swimming in the creek water and would scamper away from the creek side whenever they saw boys anywhere in the vicinity.

Even though the youth of Matawan depended on the cooling relief of their muddy, reliable creek, many of them were obligated to a summer of factory work and apprenticeship. On Thursday, July 11, now three days after the "honeymooner postcard" was postmarked at Asbury Park, a group of boys assembled to play pick-up baseball at a vacant lot near Stanley Fisher's new dry-cleaning and tailor shop. The boys, all from Matawan, were around the age of eleven or twelve. It was 4:30 P.M. and just late enough for the working kids to be set free. It was also a touch cooler than it had been an hour earlier. At bat was Charles Van Brunt, at first base was Albert O'Hara, at second was Johnson Cartan, for third was Anthony Bubblin, in left field was Rennie Cartan, in centerfield was Frank Close, and in right field was Lester Stillwell. Lester was treated a touch differently than the other kids because, at times, he would shake with unconscious and embarrassing convulsions. Lester was afflicted with the "fits," now known as tonic-clonic epilepsy. Albert O'Hara was probably Lester's closest buddy, and Anthony Bubblin was considered the joker of the group. Johnson Cartan's parents ran the local department store. His cousin, Rennie, who got his name as the shortened version of Renselear, was playing left field, and he was preoccupied by an ugly abrasion on his right leg. He didn't make much of the bruise because every time he mention how he got his injury, he was scoffed at. Rennie had been swimming at the creek swimming hole that morning and had felt a strange sandpaper-like object rub forcefully against his leg. Rennie knew that there was no piling or submerged objects in the spot at which he was swimming, and he concluded that some type of live monster was lurking in the murky water.

While Charles Van Brunt was at bat, his teammates, who included Bill Burlew, Jerry Hourihan, Leroy Smith, and eight-year-old Johnny Smith, were attempting to hunt down at least one more player. The boys were immediately in luck. Bill Burlew's older cousin, George "Red" Burlew, was just making his way by to visit with Stanley Fisher at the new dry-cleaning business. George was a tall young man who loved to fish. He could be spotted anywhere in a crowd with his bright red hair. "Hey Red," Billy shouted, "we need another player."

"I guess I could play for a while, but let me tell Stanley what's going on," George replied.

When Red Burlew shouted back to Bill, little Johnny Smith got an idea for an even better plan. Johnny was the errand boy for Fisher's tailor shop, and he knew that Stanley would probably be closing up the shop at around 5:00 P.M. "Hey fellas," Johnny shouted, "why don't we get Mr. Fisher to play as well?"

The boys all agreed they would put Stanley and Red on opposite teams and have one heck of a game. They were immediately disheartened, however, when they saw a well-dressed gentleman toting a straw hat approaching Fisher's establishment. The man was none other than Arris Henderson, Matawan's illustrious mayor. The boys thought for sure that Mayor Henderson would be a late-afternoon customer just in time to hold up Stanley. Henderson, feeling the gang's stare, waved a salute in their direction, hesitated for a second in front of Stanley's place, and, to the boys' relief, kept walking. Johnny Smith finally made a heartfelt request of Fisher inside the shop, and, moments later, Fisher's 210-pound frame appeared on the field.

Standing near the field, looking a bit disappointed, was Mary Anderson. She had gone to see Stanley only minutes before Red Burlew. Mary had apparently wanted to take a short walk by the lake with Stanley after he closed shop. Before Red interrupted them, the couple was in the midst of reminiscing about their evening trip to watch Fourth of July fireworks over New York Harbor. Stanley and Mary had made an Independence Day sojourn to the platform of the brick, fortresslike Twin Lights lighthouse in Atlantic Highlands. The lighthouse, which still stands today, is a magnificent National Historic site located atop the highest elevation on the East Coast. The location must have been breath-taking to the young couple and years earlier it was the spot for Marconi's first transatlantic transmission. The two had also been seen together at the shop and near the lake a couple of weeks earlier. The lake was a man-made body that was formed after the partial damming of the creek just above the train trestle. It would one day get its name, Lake Lefferts, after longtime resident Jacob Lefferts.

Fisher's handsome and athletic characteristics were surpassed only by his kindness and popularity in the community. Just days before, Stanley was measuring a friend named Ralph Gorsline for a new Cecil suit. The garment, which was the springtime fashion rage, was guaranteed to fit perfectly, and its sales slogan stated, "If you are not pleased with it in every respect we ask that you not accept it, not to pay one penny." Gorsline, an insurance agent for London & Lancashire Indemnity, told Fisher of tough financial times and spoke of rough goings for his new family. Fisher let Gorsline know that he would do everything he could to cut him a break on the cost of the apparel so Gorsline could make a good impression on potential customers. Gorsline, being completely cash poor, offered kind-hearted Stanley a $10,000 life insurance policy in exchange for the suit. The deal was not unfair but even Gorsline was stunned at Fisher's acceptance of the trade, since Fisher, unmarried and young, was one of the healthiest individuals he knew. Fisher's reputation was so sterling that he had the local authorized dealership for the Royal Tailors of Chicago and New York.

Fisher was about to pound a self-thrown ball when he spotted retired sea captain Thomas Cottrell walking by. Easily recognizable by his white hair and handle-bar mustache, Cottrell passed the field with a large striped bass in his left hand. "Captain Cottrell!" Red shouted, "that's some striper you got there." Cottrell was not the only retired old sailor in town. Matawan, being a stone's throw from the fishing town of Keyport and the bay, had always played host to those interested in seagoing occupations. Stanley Fisher's dad, Watson, for instance, was a retired commodore of the Savannah Steamship Line and was known to all in the New Jersey nautical circles. Red Burlew would later become captain of a sport-fishing vessel out of Brielle, New Jersey, and Newport Richy, Florida. Burlew would also become so proficient and knowledgeable at his trade that he'd be touted as one of the East Coast's most renowned big game fishing guides.

Wednesday morning, July 12, 1916, came like most others that summer: hot and humid. The heat was being blamed for multiple deaths, especially in the urban regions. Polio continued to take its toll on New York City, while New Jersey and Philadelphia were becoming increasingly concerned about further infiltration. The U.S. Bureau of Fisheries sent a communication to President Wilson and mentioned that they were virtually powerless against the man-eaters because of the vast numbers of the schooling sharks. They would attempt a hunt, however, in hopes of reducing the gross number. A dispatch from the Coast Guard noted that there were so many sharks that a patrol of cutters would be useless.

The raging battles in Europe and in the North Sea reasserted dominance over the front pages and the stories of war's hell would tell Americans just how fortunate they really were not to be involved in the fight. As the day progressed, however, Matawan, a town that to its inhabitants seemed like a world unto itself, would experience its own brand of tragedy.

Back in New York City, on the morning of July 12, Joseph Dunn, twelve, and his brother Michael, fourteen, longed to break out of their sweltering and oppressive apartment to take advantage of their aunt's bayside home in Cliffwood, New Jersey. Cliffwood Beach was the town immediately north and east of Matawan, and its southern border actually made up much of the north bank of the Matawan Creek. Joseph and Michael always loved the day trip by train to visit with their aunt, Mrs. John Murphy, who would let the boys venture to the bay area and the creek. On this day, they wanted to get to their aunt's as soon as possible. They planned to meet up with their friend, and Matawan youth, Jerry Hourihan, sixteen. To these city boys a day at the creek was a day in paradise.

Young Joseph Dunn was a diligent student at St. Stephen's Parochial School in the city, and he and his brother were close despite the age difference. Before Joseph left their home at 158 West 128th Street, he had the presence of mind to realize that his mom might worry about their safety because of the strange events that had transpired along the Jersey Shore. Joseph turned to his mom on the short row of brick steps and said, "Don't worry about us on the trip; we'll be careful."

Earlier that morning at about 6:30 A.M., Lester Stillwell left home with his father for another day at the Anderson Basket factory. Lester's mom, Sarah, never let him get away without some breakfast and a kiss on the cheek. She always gave him the same warning before his daily departure: "Come home right after work." Lester's mom was especially watchful of him compared to his two brothers and one sister, as Lester suffered from the seizure disorder. Lester's buddy Ally was also a Matawan factory worker, and he and Lester would always team up at lunchtime and scheme about games to play after chores.

Bill Burlew, whose home was actually in Keyport, was a bit older than Lester and Ally. That morning, he set off to meet John Cottrell (nephew of Captain Cottrell) for a trip to one of the local fish markets for retail purposes. Bill found his job in the fisheries industry a tad more intriguing than the year before because, this year, he got to see a lot more monster sharks that were inadvertently caught up in the pound nets. The youngest of the local workers was, of course, little Johnny Smith. While Johnny's older brother Leroy (pronounced la-roy) was headed off to a construction job in an adjacent community, Johnny was on his way to help out Stanley Fisher at the dry-cleaning establishment. On Johnny's walk past the railroad tracks, he spotted the local train engineer, Harry Van Cleaf, walking his dog. With the train schedules running early during the summer, Van Cleaf was probably the earliest riser of all the Matawan locals.

The only person who was free to go fishing was Captain Thomas Cottrell. Cottrell often spent his early morning near the Keyport bay fishing for large blues, fluke, and stripers. That morning, he made his way across Matawan's new trolley drawbridge at about 7:30 A.M. and planned to make his way back before the sun began to bake, between 1:00 and 2:00 P.M. Four construction workers were making some minor adjustments to the bridge's new mechanism when the retired sea captain made the initial trip toward the bay. The workmen, like most inhabitants of the region, knew Cottrell fairly well and wished the limping old salt good luck with the fish. The kind but notoriously crusty seaman explained that the tide was going to be high at 2:30, so he did not expect a particularly good day's catch. The lower water level was apparently not very conducive for tantalizing bait-seeking appetites early in the morning. Cottrell very rarely went out to sea anymore, but he did have access to a small motorboat docked in Keyport and sometimes joined his son-in-law, Richard Lee, on the family's twenty-three footer, called the *Skud*.

By early afternoon Charles Van Brunt came to visit Lester and Ally at the factory. The boys entered the sun-lit gabled entrance of the factory and spotted Lester pounding his hammer at a peach basket. The boys made sure that Mr. Stillwell was not within earshot and then began speaking of plans to get down to the creek as soon as work let out around three o'clock. The boys ate their bagged lunches together, and Ally shared some of his prized beef jerky with the other two. Lester told the other fellows that he wanted to go down to the creek as much as they did, but he wasn't sure if his father would let him go until he checked in with his mom after work. As Charles departed, Ally turned and chided Lester by saying, "You never swim at the deep spot anyway."

While Lester and his friends were eating lunch, William Stillwell was over at Mulsoff's Barber Shop getting a trim. Mulsoff talked politics as he snipped away. "It's sure good news that President Wilson is doin' all he can to keep us out of this one."

"His political opposition thinks he should have done a lot more since that *Lusitania* sinking last year," a customer responded. In a rare show of emotion, William Stillwell added, "It's just a shame that those innocent young men are dying over there."

For most of the week, the temperature hovered in the mideighties. Today, the mercury would spike to ninety-six. At 1:45, Lester had completed 150 peach baskets. Because of Lester's hard work and in view of the devastating heat, Lester's dad dismissed his son early and set him free to join his friends at the creek. The boys always took the quickest and quietest route to the creek. They would jump right off Main Street and dart down the short side street that ran behind the bag factory. At the foot of Matawan's Water Street, Lester, accompanied by Charles, Ally, Johnson Cartan, Frank Close, and Anthony Bubblin hung his overalls and underwear on the tree branches above the high grass next to his friends' clothing. The youngsters proceeded to play their favorite game of tag on the pilings. The chase would ultimately culminate with a dramatic dive and splash. The boys' swim in the creek amounted to a dirty, mud-water bath, but to them it was perfect bliss.

Little did the Matawan boys know that about a half hour prior to their arrival at the creek, Captain Cottrell had been making his return walk from the Keyport fishing grounds. Cottrell's day was going just about as routinely as any other that summer. Yes, it was all routine until he reached the trolley drawbridge, where he saw what would make him a figure to be referenced in all regional papers and in countless publications for the next eighty-five years.

When he arrived at the bridge, the captain was surprised to see Mother Carey's chickens (stormy petrels) resting on the railing of the overpass. The bridge workmen heard him say that he had never seen this particular variety of offshore bird so far inland. Looking down toward the flowing creek water, the captain spotted a shape that forced a double take. His aged eyes refocused, and he was frightened to realize that the heat was not playing tricks on his mind or his vision. Cottrell sighted a formidable dark gray shape, approximately eight feet in length, making its way west, up the creek, with the incoming tide. Cottrell recognized the silhouette immediately because he had seen the same kind in the open sea many times before. The object was a shark, and a large one at that. Again, Cottrell questioned the reality of the situation, but one look at the expression on the bridge workmen's faces, and Cottrell realized he was not alone in his perception of the event. Cottrell dropped his catch and dashed to bridge keeper Walling's phone.

The panicked sailor's first call was to Mulsoff's Barber Shop, where he hoped to reach Frank Mulsoff, town barber/chief of police. In a 1989 interview with Marion Smith (Johnny Smith's future wife), Marion told me of the days when she was a young resident of Matawan. I can't help but recall the gracious lady's recollections of Chief Mulsoff. The only problem with that entertaining interview was the fact that her home, which was also her family homestead, sat at the end of a long dirt drive among open farm fields. It was such a picturesque rural setting that I could barely think about sharks. She couldn't forget Chief Mulsoff. "He was the only policeman in town," she stated. "My father would go to the barber shop and I'd stay in the automobile, and he'd stop by the candy store next door for me before we got going. Whenever the teenagers would try to get into the movie theater for free, he'd get after them, you know," she said with a high pitched laugh. "He was quite a character."

Chief Mulsoff reacted to Cottrell's distress call the way most would have. The chief believed that the earlier July shark attacks had forced Cottrell to misinterpret what he had seen. In other words, he thought the old guy's soft mind was playing tricks on him. Patrons in the barber shop, eavesdropping on Mulsoff's conversation, began to laugh aloud and question whether the heat and humidity might be causing the elderly Tom Cottrell to see dock timbers masquerading as sharks. One gentleman chuckled and said, "You have a better chance seeing an elephant cooling off down there than a shark." Cottrell, bursting with frustration, knew he was not the victim of a hallucination because the four workmen on the dock had also seen the monster shark.

From the bridge, which stood about a mile and a half from the center of Matawan, Cottrell, driven by adrenaline and rage, set out in a motorboat to warn creek swimmers of the shark's presence. Cottrell made his way up the creek a good distance before running up toward Main Street, which ran parallel to the creek, to warn the boys headed down for a swim. In a freak twist of fate, however, which the well-intentioned Cottrell would never forget, he ran past the Wykoff dock area just seconds before Lester and the gang arrived for a swim. He did not get to warn them.

Cottrell's run also included a dash from store to store declaring what he had seen. Like the crew at Mulsoff's Barber Shop, the residents and businessmen of Matawan scoffed at such claims and also wondered whether the extreme heat finally had its way with the old sea captain. Cottrell even made his way by Stanley Fisher's dry-cleaning place. Fisher's comment to visitor Mary Anderson was similar to the other townfolk, "I didn't realize Captain Cottell had such a splendid sense of humor?" The townspeople simply could not believe that a shark could have made its way up a meandering tidal creek sixteen miles from the open ocean.

Shortly after Captain Cottrell's motorboat swept up the creek, Joseph and Michael Dunn, having arrived from New York earlier that day, made their way to their favorite swimming spot. The boys joined their local friend, Jerry Hourihan, at the New Jersey Clay Company brickyard docks along the north side of Matawan Creek, approximately one half mile east of the Wyckoff dock. The boys had chosen this location because the creek was deeper for diving at that dredged portion, and, most important, the dock ladder was very convenient for entry and exit. The boys did have to use some deception at the end of their walk. They knew that the superintendent of the dock, Robert Thress, would not like it if he saw them playing in front of the loading zone. When he wasn't looking, they crept like commandos right down the dock ladder.

At the Stillwell home, William Stillwell's heavy steps brought Sarah Stillwell to the front porch. Sarah was immediately surprised that

Lester had not joined his dad on the trip home. William calmed her and explained that he let the deserving boy out of work early to go for a swim.

At a little after 2:00 P.M. at the Wyckoff dock area, Lester and his friends were playing in the muddy water, unaware of the commotion in town. Ally O'Hara, swimming a bit closer to shore than the other kids, suddenly felt a sandpaper-like object graze his leg. He immediately trained his eyes beneath the tea-colored water and saw what looked like the tail of a huge fish or a seamonster. Lester was swimming closest to O'Hara, but before Ally could muster the words to describe what he had just experienced, Lester wriggled his thin frame toward the deep spot. At the same time, Johnson Cartan and the other boys spotted what appeared to be an "old black weather-beaten board or a weathered log" bob the surface of the water. Boisterous Anthony Bubblin, who had now climbed on to one of the higher pilings, announced that he was going to attempt a back flip. By now, Lester was proud to sustain buoyancy on his back at the deeper hole and exclaimed, "Hey fellas, watch me float!" Most of the boys were still focused on Anthony's lofty attempt and burst into laughter when the boy created a huge splash with his less-than-perfect dive. Before the airborne water had a chance to rejoin the brown creek, the boys heard a short screech and an even greater splash behind them. The children were momentarily entranced by what they originally believed to be a dark old plank surge toward Lester. Then they saw the dorsal fin and tail fin of the shark and in unison shouted, "Lester's gone!" As the phantom engulfed Lester's slight upper body, Lester's mouth filled with water. The fear-frozen boys saw at once that the beast was not all black but had a white underside and gleaming teeth. Poor Lester struggled briefly amid the reddening water, gurgled a scream once more, and was dragged beneath the surface.

The boys scattered from the creek like flies evading a swatter. The only one to turn and stop, if only for a split second, was Ally. Some of them ran toward the Fisher bag factory (no relation to Stanley), which bordered the creek. Others raced up the steep dirt path at Water Street and headed for Main Street, screaming, "Shark! Shark! A shark got Lester!!!" Amid the frantic commotion, none of the boys remembered to clothe themselves.

In 1916, Johnson Cartan was a fourteen-year-old swimming companion of Lester Stillwell. Cartan later went on to become a prominent local attorney, and he would spend his entire life living in Matawan at 92 Main Street. In 1916, 92 Main Street was the Cartan Department Store (est. 1890), which moved to 121 Main Street in 1930. In 1985, as I sat down in Mr. Cartan's living room with him and his wife, I silently marveled at the fact that so many 1916 witnesses had lived to such ripe ages. The soft-spoken Cartan told me, "We were swimming over near what we called the Wyckoff dock. We swam in that water when it was so dirty. It was so dirty that today I wouldn't even stick my finger in it. Every day we swam there; every day, you know. We didn't think it was a shark. I thought it was an old weather-beaten board. When you're young, you never think of a shark coming up a little stream like that, you know? Of course, they didn't believe us that there was a shark."

By 1990, Anthony Bubblin's daughter, Annette Baker, still remembered the account that her late father had told her of the attack on young Stillwell. "He actually stood there petrified, and the shark kind of like swam around him. The faired-skinned boy was attacked, and my father was always quite a storyteller and quite a character, but this story had been brought up so many times that there must have been some amount of truth to it. Otherwise, stories seem to grow over the years, but this one always came out pretty well the same. When you're younger, you only seem to remember those things that are exciting."

The panicked boys ran into town. They passed Stanley Fisher's dry-cleaning establishment crying out that a shark in the creek had gotten Lester. Stanley commented that Matawan must be going shark crazy, but

Mary Anderson, inside Fisher's store and closer to the store's front window, turned to Stanley and remarked that the boys were nude and that the concern in their eyes and their voices seemed quite real. The errand boy, Johnny Smith, walked into the shop just as Mary made her comment. Johnny was carrying some clothespins and suit buttons, and Stanley told the boy to place the items by the register. Stanley's routinely jovial demeanor abruptly turned serious and pensive. "Didn't those boys say that it was Lester that was attacked by a shark?" Fisher asked.

Seventy-four years after that fateful day in Matawan, I was able to track down Johnny Smith's older brother, Leroy, residing with his wife of sixty years on the east coast of Florida. Mr. Smith was most kind to draw for me a map, which depicted the exact location of all the major stores during the 1916 period. He drew in the location of Stanley Fisher's home, the dry-cleaning shop, the Smith homestead, and assorted other spots which would never be indicated by standard atlases or documentation. He was also kind enough to send me an announcement of his sixty-fifth wedding anniversary. In his late nineties, Mr. Smith remembered Stanley Fisher with warmth and admiration. "Us kids used to play ball out in the field between our house and where Stanley Fisher had his establishment," Smith stated. "He used to come out and play with us kids once and a while. . . . Oh, he was a very nice man, he was well liked, and he was real good lookin', a big man. He was friendly to everybody."

John Applegate later became Matawan's mayor and had some interesting memories of that time and of Stanley Fisher, as well. Applegate recalled, "I can remember him by the fact that as kids growing up . . . What the hell, there were only twelve hundred or a thousand people then. . . . The young men who were good at something, like baseball or something like that, were well known to the kids. I knew him in that sense."

As the boys ran shouting past his store, baseball was the furthest thing from Fisher's mind. He was thinking instead of Lester's epileptic condition and whether the boy could truly be in trouble. In a commanding voice Fisher exclaimed, "Johnny, tend the store till I get back." As the towering blond giant darted out the door, Mary Anderson shouted, "What about the matter Captain Cottrell mentioned about a shark?" Practically running Mary down on his hasty exit, Fisher did not waiver, but turned to her and said, "A shark here? I don't care! Lester's got the fits. If we don't get to him soon, he'll be finished." Outside the store, Fisher grabbed Red Burlew, who was just passing by, and middle-aged carpenter Arthur Smith. He asked the men to join him at the creek.

When they arrived at the Wyckoff dock, the brown water was stained red with blood. The three men first utilized a small rowboat to attempt to locate Lester. It was now close to a half hour after Lester disappeared, and the men were resigned to locating the corpse instead of managing a rescue. Stanley commented on the strange redness of the water, and Burlew wondered aloud whether the boy might have hit his head forcefully on a piling or a rock. Smith suggested that the boy's fit (seizure) might have caused some blood-tinged vomiting. While the men poked the depths with long rods for an indication of the body, other concerned citizens, including Asher Wooley, began to arrive. Wooley brought chicken wire and grappling hooks from his hardware store, realizing that some form of netting would likely help in the search. Wooley and the other men hung the fencing under the nearby train trestle overhang and across the banks near the dock pilings. By doing so the men hoped to entrap the boy's remains and prevent them from being taken with the current. What they did not realize was that they also could be entrapping a killer shark.

With virtually the entire town making its way down to the dock, Fisher, Smith, and Burlew became even more driven to locate the poor victim. Ally O'Hara returned to the creek side with his dazed companions. Johnson Cartan corroborated the account by informing me that, "George Burlew, Stanley Fisher, Arthur Smith . . . They were all there. They were older than us, of course, they didn't believe us that there was a shark." The brave men borrowed swim tights and donned them behind a dense tree. They began making multiple dives for Lester. Arthur Smith was making dives closer to the dock when his abdominal area was painfully scraped by a moving object. As the fifty-one-year-old Smith struggled to gulp more air, he noticed blood seeping through his shirt from a nasty abrasion.

Over forty years later, the senior author of *Shadows in the Sea*, Thomas B. Allen, interviewed an elderly and blind Arthur Smith. During the conversation, Smith touched the residual scar on his stomach region and questioned why his life had not been claimed during the incident.

At the opposite side of the creek, Fisher believed that Lester's body was resting at a deep hole, and he was intent on surfacing with the boy. Fisher dove to the bottom and thought he saw a log rolling on top of the boy's body. At the same time, Red Burlew, who had joined Fisher on the far side of the creek, encountered an unusual swirl of water just as Fisher surfaced. Burlew also noticed that Arthur Smith was grimacing in pain near the dock pilings. When Fisher surfaced, feeling optimistic about recovering Lester's remains, he noticed the look of dread on Red's face and the nature of distress that overtook Arthur Smith. Red had clearly begun to lose his resolve to recover the child, and both he and Stanley decided to call it quits. Red said, "I'm cold," and Stanley responded, "Me, too."

Along the south side of the creek, near the old dock, the concerned Matawan residents were now densely assembled. Among the crowd was Johnny Smith, who was accompanied by Mary Anderson. One glance at Mary represented to Stanley that there was everything to live for and he shouldn't do something foolish. He could be risking a fulfilling future.

But, before Stanley and Red made their way back across the narrow creek, Fisher noticed a tearful Sarah Stillwell and her husband standing on the bank. In an instant, Fisher decided to make one last dive for the boy. Fisher, as brave as he was well liked, filled his lungs to capacity and made the dive through the brown soup, all the way to the bottom. Again, he saw what he thought to be a log, now moving frantically about the boy's corpse. He tugged at Lester's legs and grabbed him around what remained of his torso. Ultimately, Fisher broke the surface with the boy's thoroughly white corpse. A gasp from the crowd was heard as some of the spectators perceived the tattered remains of Lester under Stanley's arm. Some parents shielded the eyes of their young children. Asher Wooley was just about done securing the chicken wire under the train trestle when he heard the crowd's response. He looked over in Stanley's direction. As Fisher attempted to plant his feet near the muddy slope, in waist-deep water, the unthinkable happened. The horns of the devil were no longer concealed. Fisher was violently hammered on the right thigh. In an attempt to maintain balance, Stanley was forced to let Lester slip from his grasp. The crowd could not muster a word or a breath. Fisher shrieked, "He's got me! The shark's got me!" and frantically punched and kicked the phantom menace with every ounce of his athletic strength and one simple objective: to avoid meeting Lester that day in the great beyond.

Moments earlier, Arthur Van Buskirk, a deputy at the Monmouth County detective's office in Keyport, and George Smith, a Freehold detective, had heard about the commotion in Matawan from Keyport residents. The men made their way in a motorboat toward the Wyckoff dock just as Fisher's struggle began. They later reported that Fisher fought the shark like a madman and used every fiber of his muscular body to ward off the monster. Fisher was brought under twice, and spun around but Van Buskirk directed his boat toward the widening red stain and furiously shouted and slapped the shark with a boat oar to get it to release Stanley.

In the mid 1980s, I interviewed Bill Burlew, the nephew of Red Burlew, about the events at the creek on that unforgettable day. Mr. Burlew, in his eighties at the time of our conversation, was not at the Wyckoff dock that afternoon; he was closer to Keyport doing work for the fisheries industry. I remembered how enthralled I was to hear Mr. Burlew recount the story his uncle George (Red) had conveyed about the creek incident. During the course of our talk, I noticed that Mr. Burlew kept referring to his uncle's thoughts in the present tense. I assumed that, regardless of how old a nephew or an uncle grows, the nephew would still consider the uncle as a vibrant older friend. With Bill Burlew already in his mid-eighties, I assumed that Red had passed on a decade or two before. Still, I was compelled to ask, "Your uncle George is not still living, is he?"

"Why yes," he said, "he's about ninety-five now, living down in Florida with his sister."

Telling the story of the Matawan Creek attacks was not alien to Red Burlew. As a sea captain, the topic would frequently arise during overnight fishing trips or when formidable sharks were hooked. Burlew, in fact, won a journalistic prize in the 1930s when he offered Floyd Gibbons, the famed World War I writer, his account of the 1916 events.

In 1985, seventy years after the attacks took place, the ailing Red Burlew, not more than six months away from passing on himself, remembered the afternoon vividly. Burlew informed me that, after several dives for Stillwell, he was resting with Stanley Fisher near the creek's meadow bank, opposite and just west of the Wyckoff dock. It was in that location where he first encountered the shark. "We knew [Lester] was dead, but we wanted to get him out. And the water was so thick, you know, you couldn't see anything. You'd get the wind outta ya, and you'd have to come up [for air]. I was standing in the water right by [Stanley]. He was standing back off the little shelf over there. I saw the faint [shadow] of him, you know, and the water was kind of thick, you know. They're very scary. [The shark] turned away from me; that's how I came to know him. After I saw the faint of the shark, I told Stanley that I was cold and I was gonna get out. He said, 'Me, too.' But he didn't have a chance to start across, the shark just came up and took him! Boy, I went for that dock!"

After a violent struggle, the shark finally released Fisher. Fisher made his way to the slope near the dock. Though still alive, Stanley was not fully aware of the severity of his injuries. He was standing, slumped at the dockside, clenching his right thigh. The phantom shark had thrown the crowd into such a state of delirium and fear, no one was willing even to jump into the water to assist Stanley. A rope was extended to him, and he was pulled out by the motorboat, assisted by his own efforts. As his body was pulled from the water, his right thigh was revealed, stripped of ten pounds of flesh with bone exposed and blood gushing and spurting in all directions.

In my analysis of the documentation and eyewitness accounts, I have made an issue of whether Stanley ever actually retrieved Lester's corpse from the bottom of the creek. To me, the circumstances revolving around the assault on Stanley Fisher are critical to a better understanding of such bizarre multiple attacks. My scrutiny of the initial retrieval of Lester was rooted in the limited visibility of the creek water as well as the fact that a couple of the witnesses to whom I've spoken never actually saw Stillwell's corpse in Stanley's arms. They also never heard Stanley report that he had seen the boy. Most newspaper reports, however, claim that Stanley did find Lester and that witnesses heard him say so. Certainly the chaos of the event and the differing angles of vision could account for the discrepancies in the reports. In my conversation with George Burlew, he did describe the shark's agility and ease in jerking Stanley in any direction it chose. Burlew also corroborated the three chilling words that Stanley voiced when he first saw his massive thigh wound. Burlew said, "I saw that shark clearly when it took him and spun him around, it brought him around twice. It took a big piece out of him, and the blood circulation came out. Oh, it was just terrible. . . . The only thing I heard him say [when he looked down at his leg] was 'Oh my God.'"

As Stanley was hoisted onto the old wharf, a forceful stream of bright red blood pulsed skyward. The crowd gasped, and several woman fainted. Fathers again shielded the eyes of their young children, and Mary Anderson grabbed young Johnny with a crushing hug. Onlookers shouted for the local doctor, and little Alfreda Metz, standing in the distance with her dad, commented, "I believe a crocodile must have bitten Mr. Fisher." The girl's soft-spoken remark was followed by an impressive clatter generated by the displacement of chicken wire at the eastern edge of the dock. The hastily secured fence was hurled down as if a Mack truck had rammed it flat.

As Stanley lay moaning on the dock, Dr. George C. Reynolds fastened a rope around his upper right thigh in an attempt to slow the pressured blood loss. Matawan physician Dr. Reynolds was summoned and was enlisted to treat a type of wound that was alien not only to his extensive experience but to any American physician until just days before. The doctor described the wound as a wide, jagged laceration, measuring approximately eighteen inches, spanning from below the hip to just above the knee. Dr. Reynold's original estimate of the wound length was later determined to be fourteen inches, not eighteen. The amazed physician also noted that approximately ten pounds of tissue had been removed and the remaining flesh appeared to have been racked with dull knives. The femur was exposed and scratched, and the femoral artery severed.

Long-skirted mothers ran frantically along the creek bank to alert any of the children who might still be swimming. Captain Cottrell did not gloat about the fact that the townsfolk dismissed his original warnings. He simply resumed his motorboat patrol to warn any remaining creek bathers.

At the Wyckoff dock, Stanley Fisher, who was now fully cognizant of the horror, writhed in pain. He asked the doctor if there was much tissue loss, and the doctor said, "Not much." Mayor Arris Henderson and Dr. Reynolds leaned over Stanley and did all they could to address his concerns; it was here, according to other accounts, that Stanley first reportedly stated that he had seen the body of Lester Stillwell on the bottom and had ripped it from the jaws of the shark. Stanley asked the doctor for something to alleviate his pain and then asked whether his leg would have to be amputated. The doctor responded, "It's not so bad." William Shepard, a close friend of the Fisher family and a part-time assistant at the dry-cleaning shop, leaned over Stanley to tell him about the arrangements to get him to the hospital. Fisher told Shepard not to take any new suit orders until he returned and asked Shepard to make sure that his sister was in good hands if anything should happen to him. At this point, the desperate men improvised a stretcher made out of planks and transferred Fisher up the embankment to the train station.

Mayor Henderson was informed that it would be at least two hours before the next train would enter the station; he looked to Harry Van Cleaf, the local engineer, who was at the creek side with his inquisitive daughter, Mildred, for some suggestions. Van Cleaf arranged for the next available train to skip as many stops as possible on its way to Long Branch and the Monmouth Memorial Hospital.

Mary Anderson was escorted to a friend's home, essentially catatonic from the day's events. Neither she nor any other Matawan citizen ever imagined that their peaceful farm town would wake that morning to find an exotic monster from the tropics in their own backyard. A routine summer Wednesday turned into a nightmare that Matawan and the world had never seen before. And the horror was not over yet.

The attacks on Charles Vansant, Charles Bruder, and two of their own should have told the people of Matawan that science, logic, and experience meant nothing to the Jersey man-eater. This was not a creature that obeyed the laws of nature and predictability.

At the brickyard dock, approximately a half mile away from the Wyckoff dock, the Dunn boys, Jerry Hourihan, and two other friends were still swimming, immune to the ravages that had struck a half hour earlier. Suddenly, the boys heard faint warning cries, the muffled sound traveling over the high reeds and grass. All they could make out was the word "shark." They froze in unison, and each turned to measure his distance from the dock ladder. Two of the boys got to the top of the ladder and leaped from the rungs. Michael was next up the ladder, then Jerry Hourihan. Joseph was not only the farthest away, he was also the youngest. He finally made his way to the lowest subsurface rung of the ladder and grabbed for a piece of Jerry's tights to help hoist him upward. Michael's and Jerry's silent jubilation at having reached dry land was abruptly shattered by a most disconcerting shriek and splash. They had not yet stopped hyperventilating from their dash up the ladder when they glanced over the side of the dock and saw what they most dreaded. Joseph's head was bobbing just above the surface and was now ten feet from the dock. It appeared he was being tugged. The other boys on the dock quickly ran for help.

Joseph, who felt a piercing crunch to his left lower leg, looked down to see a dark shape and then a diffuse explosion of reddening water. He didn't remember much after that.

The older boys hoped that Joseph had simply lost his footing and would make his way back, but as the water oozed red with blood, they knew the situation was serious. Joseph was being tugged under the surface like a thin cork. Without more than a second to think, Jerry and Michael created a human chain to get Joseph to safety. Joseph was halfway to the point of no return when his brother Michael jumped back into the water and decisively grabbed his hand. With every ounce of youthful strength and blind determination, Jerry and Michael struggled to pull Joseph free. Just as the heroic boys were beginning to tire, the brickyard superintendent, Robert Thress, rushed up behind them, seized Joe's forearms and, with a mighty jerk, lurched the young boy free. Jacob Lefferts, a thirty-five-year-old, well-known Matawan resident, was fortunately in the vicinity fishing. Lefferts selflessly dove fully clothed and headfirst into the treacherous waters to make sure Joe got safely to the ladder.

Just as Jacob Lefferts helped the boys get Joe Dunn to the dock ladder, Captain Cottrell cruised up in his motorboat. Before the red water calmed, Lefferts had apprised Cottrell of the crisis. Cottrell quickly bundled the boy and transported him and his brother to the crowded Wyckoff dock area. There, Cottrell hoped, the Dunn boy would receive speedy care from the crew that was assembled for Stanley Fisher.

As Cottrell's buzzing craft approached the dock through the blood-tinged creek, the residual crowd was dismayed to realize that there had been yet a third attack, and their emotion turned to anger. A robust onlooker shouted to Asher Wooley, "Hey, Asher, got any dynamite at that store of yours?"

"Sure do," the thin but determined Wooley responded. At that moment, the frenzied crowd was set on vengeance rather than mourning. They were no longer passive to the calamity that had been set upon them by the grisly monster from the deep.

Marion Smith, just a small girl at the time of the frightening scenario, vividly recalled her parents' interest in the commotion. After we discussed the role her husband (Johnny Smith) played in the Fisher saga, she described her arrival at the scene near the creek and the Matawan station: "I remember the incident and all the commotion in Matawan. People going down Ravine Drive and trying to see where it had happened. I recall my father stopping and picking up my aunt and uncle on Main Street and going down Water Street toward the creek. At that time there were no local hospitals, and I recall them talking about getting Stanley Fisher on the train to Long Branch. He was pretty well torn apart and bleeding. I think my uncle even jumped on the train when it took off, he was that type of person. In those days there was no TV or radio, so anything like that was a big thing in their life."

The late Mildred Fisher (no relation) was also a young girl in Matawan in 1916. She recalled how her dad, Harry Van Cleaf, held her by the hand as he confronted the situation. I had the pleasure of meeting Mildred at the Clifton home in Matawan in the late 1980s, and, besides providing me with her priceless account of the events, she also identified several old-time residents from the classic creek-side photos, including narrow-faced Asher Wooley. "I remember the lane, the road that ran right down to the crik," she said. "And at that time it was a wide crik, and some boats came in there. There was a lot of excitement, so everybody in the neighborhood ran down. They said somebody had drowned. So we went down. There were a lot of older men there and boats trying to fish the bottom to get the boy out. His name was Stillwell. We all stood on the dock, and this one man, Stanley Fisher was his name, says, 'I'm going in.' So he jumped in and he found the body, but he came up, and when he came up, he said, 'I've got the boy but the shark's got me!'"

With Stanley at the Matawan station fighting off unconsciousness, Joseph Dunn was carried to the Fisher bag factory, which bordered the creek. Dunn was tended to by another Keyport physician, Dr. H. S. Cooley. Cooley found that "the front and side portion of the boy's lower left leg was cut into ribbons from knee to ankle. The bones were not crushed and the main arteries in the calf of the leg were not cut." With Stanley Fisher's chances, as he waited for the train, becoming dimmer by the minute, the doctor and the other concerned residents chose a different mode of transport for Joseph. His less severe injuries and his relative hemodynamic stability (less blood loss and more stable blood pressure), made him a better candidate to survive the bumpy car ride to St. Peter's. Between the primitive shock absorption of the automobiles and the dusty potholed roadways, the ride would not be an easy one, especially when time was of the essence. Although to this point, no 1916 attack victim had been in any condition to tolerate such a transport, the doctor eventually had the boy carried to a motor car, where he could personally transport him to New Brunswick and the St. Peter's Hospital.

Regardless of Dr. Cooley's appraisal that Dunn's major lower extremity arteries were intact, Dr. Reynolds, the physician who tended to Stanley Fisher, told Cooley that Dunn would certainly die. Reynolds and Cooley both knew of the two previous New Jersey shark fatalities and obviously knew of the demise of young Lester. Additionally, Dr. Reynolds was particularly astonished by the victim's deadened perception of initial pain. These grim facts led Reynolds to theorize that sharks transmit a lethal venom in their bite that first numbs the wound, and that the victim would inevitably die as he succumbed to the toxin.

Joe Dunn, oblivious to the dismal chances that the town doctor had given him for survival, smiled at the mass of onlookers. While the concerned crowd hovered over Dunn, a throng of reporters and photographers could be seen making its way from town hall. A Keyport man, Edward Dominic, had arrived with his 1913 Buick touring car and volunteered to transport Dr. Cooley and the injured boy to the hospital. One hungry reporter asked Joe, "How'd it feel to be bitten by the Jersey man-eater? What happened?"

As the car door was closing, Joseph explained, "I was about ten feet from the dock ladder, when I looked down and saw something dark. Suddenly I felt a tug, like a big pair of scissors pulling at my leg and bringing me under. I felt as if my leg had gone! I believe it would have swallowed me."*

Michael Dunn was quoted as saying, "I heard Joe cry, and saw him going under. I rushed out to him with Jerry Hourihan and we seized him just in time, but we could feel the shark tug at him as we pulled him away and swam toward shore." Joseph also explained that he felt the shark scratch his leg just prior to the actual bite. "[It] felt as if [the shark] was trying to get my entire leg down its throat."*

Just before the automobile drove off, another reporter approached and asked Joseph, "Will you tell me your full name?"

"I should say not," replied the usually polite youngster, "you would tell my mother."

Then the astounded reporter asked, "Where do you live?"

"In New York City," Joseph answered, "but I won't tell you the street." Without the ease of universal telephone communication (and cell phones) that we now take for granted, the badly injured boy was concerned about alarming his mother before he had a chance to tell her that he was okay. The frustrated news reporter ended up getting the information from Michael Dunn.

---

*Asbury Park Press.

Perhaps because Stillwell and Fisher were two of Matawan's own and the fact that the attacks occurred in the shadow of Main Street, the shark incident involving Joseph Dunn would, in decades to come, go generally unremembered as a tragedy of that July day. In 1936, a twentieth-anniversary commemorative map of the attack location was published by the Matawan Historical Society. The map does include the brickyard docks in nearby Cliffwood, but it fails to make any notation of the gruesome attack on Joe Dunn. The late Jerry Hourihan, however, the only Matawan resident to join the Dunn boys, never forgot that day and his encounter with the shark that almost swallowed his friend. In 1989, I spoke with Jerry's son, Jerry Hourihan Jr., still a resident and a municipal employee of the town. As Mr. Hourihan and I stood at the site of the attacks on Stillwell and Fisher, he proudly discussed his dad's involvement with the Joe Dunn rescue. Mr. Hourihan noted that it was only after the creek was narrowed near the train trestle in a million-dollar fill project in 1953 that his dad spoke of the Dunn attack. His dad drew some blank looks from those who had no idea a third attack had occurred. "My dad had mentioned that it was the forgotten attack," Hourihan remembered. "There was never too much mentioned or made of it, and some of it was not even discussed until sometime after all of these trees were cleared away."

By 5:06 P.M. on Wednesday, Stanley Fisher was finally loaded onto the train to Long Branch. The plan was to get Fisher to the operating room where he would have his leg amputated. Stanley was placed gently in the center of a train car aisle, and a makeshift bed was prepared. The conductor entered with Harry Van Cleaf and told Stanley's attendants that he would fire up the engine to top speed whenever possible. Weak, and semiconscious, Stanley still had the presence of politeness to say to the conductor, "Thank you, sir."

At St. Peter's Hospital, a previously quiet corridor was now jammed with reporters. After a lengthy assessment by the general surgeon on call, Dr. R. J. Faulkingham, a statement was prepared by the hospital's nuns for the press on the condition of Joseph Dunn. The surgeon reported that the boy's left Achilles tendon was not severed, but the calf muscles (gastrocnemius and soleus) were severely lacerated, and the smaller bones of the ankle were pierced as if they had been drilled by a sharp instrument. Although many newspaper accounts (and later shark attack narratives) incorrectly reported that Joe Dunn would likely lose his leg to amputation, Dr. Faulkingham was optimistic about saving the boy's leg. Faulkingham, however, felt that Joseph would certainly need a skin graft.

While the surgeons at St. Peter's were preparing for surgery, the people of Matawan were preparing for an operation of their own. The Matawan residents believed that the creek monster would return to the site of the first two attacks at high tide. The incensed crowd lay in wait with a hefty supply of dynamite, shotguns, harpoons, rifles, garden hoes, ice picks, axes, pitch forks, and even hammers.

Stanley's father, Commodore Watson Fisher, received a telegram from a Western Union delivery boy at his daughter's home in Minnesota. The wire was sent by Mayor Henderson and said simply that Stanley had met with a freak accident and was gravely ill. In the Anderson home that evening, distraught Mary prayed forcefully for a miracle.

As promised, Fisher's train made it to Long Branch in record time. Stanley was wheeled into the operating room at Monmouth Memorial Hospital at 5:30 P.M. A bearded visiting surgeon, Dr. Edwin Field, barked orders to the nurses and surgical assistants. The doctor inserted an intravenous line for saline and pressors. His examination revealed that blue blotches and streaks covered assorted areas of Fisher's body; as a result he, too, speculated that a toxin had been released with the shark's bite. To the nurses, the physician shook his head and remarked that he had cared for scores of men during the Spanish-American War but had never seen a wound quite like this one. Fisher was mouthing words to Dr. Field, and the surgeon, realizing that Stanley could not generate the volume to be heard, placed his ear near Stanley's mouth.

Once again, accounts tell how Fisher told Field he pulled Lester away from the shark. "Doc," Fisher muttered, "I found the boy on the bottom; I got Lester away from the shark. Anyhow, I did my duty." After Fisher breathed his last, Dr. Reynolds arrived at the hospital and told the staff how he had pleaded with Fisher not to dive for the corpse. Reynolds was informed of Fisher's last words and added, "Yes, I saw Stillwell under his arms before the shark came at him. He described it like a dog going after a bone. He felt it was a task that was part of his obligation as a human being. That was Stanley."*

Despite all efforts, Stanley Fisher expired at 6:35 that evening. It was far too late to reverse the onset of uncompensated hemorrhagic shock. When requested, Dr. Field sent a report of Fisher's death to the district superintendent of the Coast Guard, John S. Cole, at Asbury Park. In a letter, dated July 17, 1916, Dr. Field wrote:

> W. Stanley Fisher was admitted to Monmouth Memorial Hospital, Long Branch, N.J., July 12, 1916, at 5:30 P.M. Was injured at the old pier in Matawan Creek, Matawan, N.J., while endeavoring to recover the body of Lester Stillwell, who had been killed by a shark. Fisher dived into the water and had found the body, when he was seized by the right thigh. The muscles were torn out. Fisher was taken out of the water by friends nearby in a boat, but bled profusely before medical assistance could be summoned. A tourniquet was applied and he was hurried to the hospital. Condition on admission: suffering from shock and pulselessness from loss of blood. Did not rally to stimulants and saline transfusions. The outer side of the right thigh was denuded from three inches below the greater trochanter [lateral hip point] to two inches above the knee, all of the muscles and tissues being completely removed, and only a third of

---

New Brunswick Times.

the muscular tissue on the inside of the thigh remaining. Bone not injured. Died at 6:35, one hour after admission.

Dr. Reynolds did get to Fisher fairly rapidly at the dock. According to reports, he wrapped a rubber type of band around Fisher's thigh at first, but the band broke. He then used a regular tourniquet. Whether the treatment was helpful or not was not an issue. A wound in that region and of that severity without modern transport or portable intravenous systems would be very difficult to treat successfully. Because of the limitations of the era, ensuing disastrous sequelae were inevitable.

I am not certain how to clarify the evidence of blue blotches and streaks found on Fisher by the operating room staff at Long Branch, but one of three possible explanations may suffice. The easiest explanation is that the hospital staff may not have even found such blue discoloration, but rather the press reporters may have taken liberties in embellishing an already sensational story. In my self-published book *In Search of the "Jersey Man-Eater"* (1987), I suggested that a condition called thrombosis (intravascular clotting) may have occurred to create the blue marks. Such a phenomenon is seen in the progressive stages of shock when blood flow becomes sluggish and the end products of metabolism accumulate. But a third explanation relates to the medical staff's ignorance of the shark's texture and Stanley's physical impact with the beast. The tough, sandpaper-like, muscular hide of the shark, impacting a 210-pound man, would have inevitably caused at least a few major contusions (bruises) to areas of Stanley's battered body.

In the end, Stanley Fisher's death certificate would sum up the vital statistics of his life and death. "Born April 12, 1892; Occupation: Clothes cleaning & Pressing; Birthplace: New York; Mother's maiden name: Celia S. Waters." Since Stanley's parents were still in Minnesota, Johnny Smith's mother, Mrs. Eugene Smith, provided the personal information. The medical portion of the certificate revealed the cause of death: "Bitten by shark, Matawan Creek, N.J., below hip, bleeding to death upon reaching this hospital for treatment." At 10:00 P.M., the Fisher family in Minneapolis were anxiously awaiting word on Stanley's condition. A final brief telegram was delivered. The message: "I regret to inform you that Stanley has died as a result of his injuries, Arris Henderson." The Fisher family did not yet know exactly how Stanley died.

The news of Stanley's death spread rapidly along the telephone lines and the rail system. The Matawan residents now added their own grief to that which was spawned in Beach Haven and Spring Lake. Back at the creek, an even angrier crowd of Matawan citizens descended the embankment, seeking vengeance for the life of an innocent boy and one of Matawan's most popular men. The mob was intent on making revenge answer the question of why a sea wolf had dropped terror on their quiet rural town.

$100 Reward!

The above reward will be paid to the person or persons **KILLING** the **SHARK** believed to be in Matawan Creek.

In the event there is more than one shark killed, a pro rata sum will be paid for killing each shark.

Arris B. Henderson,
Acting Mayor of Matawan.

Bounty notice posted by Mayor Henderson.

CHAPTER 4

In the Wake of the Terror

By nightfall, there were no less than fifteen nets lining Matawan Creek's sinuous passages. Underwater blasts of dynamite could be heard for miles around. Some even equated the voracious creek-side activities with that of the California gold rush. On the steps and porch of the Old Homestead Hotel, men sat with sticks of dynamite and rifles. The explosives were meant to destroy the murdering shark and halt its reign of terror once and for all. Residents also believed that the underwater concussion would at least help bring Lester's corpse to the surface. Townspeople did not hesitate to sacrifice a night's sleep to get their results.

Matawan residents suddenly had another form of incentive. Just hours after the creek attacks, Mayor Henderson contacted a printer from the *Matawan Journal* and designed "wanted" posters. The poster was nailed to every tree and every telephone pole available. It read: $100 REWARD—THE ABOVE REWARD WILL BE PAID TO THE PERSON OR PERSONS KILLING THE SHARK BELIEVED TO BE IN MATAWAN CREEK. IN THE EVENT THERE IS MORE THAN ONE SHARK KILLED, A PRO RATA SUM WILL BE PAID FOR KILLING EACH SHARK. ARRIS B. HENDERSON, ACTING MAYOR OF MATAWAN.

The next morning, based on the events in Matawan, the shore regions of New Jersey knew that the ocean bathing areas were once again (or, perhaps, still) at risk. In response to that looming disaster, the Monmouth, Ocean, Atlantic, and Burlington County congressional district offices in Washington were inundated by letters demanding federal aid to safeguard bathers.

Without a body to provide closure to Lester's parents, the new sunrise brought the chilling calm at the eye of the hurricane to the Stillwell home. The morning rays barely piercing the trees, Sarah Stillwell slowly rocked in her favorite porch chair. All of the reports pointed to the assumption that Lester was completely eaten by the monster creek shark. Lester's mother struggled to reason that if she never saw Lester's mangled body, perhaps she could think more pleasant thoughts of his death. Was Lester torn to pieces or was he at the bottom of the muddy creek, she wondered. But for now, the Stillwells waited in brutal contemplative agony.

William Stillwell knew that he must face the day and the future with the same type of stoicism by which he'd always lived. Without any trace of emotion, he carried a heavy load of uncut firewood to the chopping block at the side of the house. His chore should remind us that Matawan was not a tropical or subtropical village. It was not normally subject to the assaults of fierce exotic creatures. It was a town that had always lived in harmony with the classic quartet of seasons, including a white winter.

At Ally O'Hara's somber home, Ally was not himself. He had lost a best friend, and, to make matters worse, no one even knew where the body of his pal had gone. "Is Lester in the belly of the beast at this moment?" he must have wondered. Couldn't that have been me as easy as Lester? Why him, and not one of us other fellas? Ally's mother was empathetic to his plight and allowed him to stay out of work as long as he told his employer. She, like all Matawan mothers for the next generation, demanded, however, that he never swim in the Matawan Creek again.

A glance at any newsstand in the country would reveal that Matawan had catapulted to the national limelight. The town was the object of widespread curiosity and sympathy. The shark was the object of wide-spread infamy. The foreign battles of World War I could not hold a candle to the bizarre excitement surrounding the rampaging man-eater from the sea. The regions of New Jersey and the country that were not directly bordering the ocean had felt immune to the entire shark plague. Now, with inland Matawan being violated, no one felt safe.

The hungry press was writing with pens already red hot from the previous two attacks. This was a horror novel being played out in real life, the ending yet unknown. Striking headlines like the following were the norm: FUTILE EFFORTS TO CAPTURE SHARK SLAYER OF LAD AND MAN: DYNAMITE AND NETS FAILS TO NAB MAN-EATER OR RECOVER BODY OF BOY KILLED AT MATAWAN. WOULD-BE RESCUER FATALLY BITTEN ATTEMPTING TO SNATCH YOUTH FROM JAWS OF MONSTER: WILL SAVE THE LEG OF OTHER YOUTH INJURED." *(Asbury Park Press)* Other approaches to the headlines captured the essence of the shark hunt and made Jersey residents feel as if they were living in a gunslinger's paradise. One headline from the *Chicago Tribune* read: ARMED POSSES COMB COAST TO SNUFF OUT MAN-EATING SHARKS; $100 REWARD OFFERED FOR CAPTURED MONSTERS DEAD OR ALIVE. Across the Atlantic, however, the British prime minister and the Parliament were more concerned about discussing methods that might compel the United States to join in the war. An adviser to the British officials informed the staff that the United States might be preoccupied with more pressing national problems. The man held up the morning *London Times,* which revealed that the United States was busy with a war of its own. The headline describes: DANGEROUS SCHOOLS OF SHARKS TERRORIZING THE COAST OF THE U.S.

Three individuals who could not afford to be caught off guard by any further shark-inflicted developments were President Woodrow Wilson, Dr. John Nichols, and Dr. Frederic Lucas. On the morning of July 13, President Wilson, who had yet to announce the location of his summer residence, received a telephone call from the haggard Treasury secretary, William McAdoo, in Spring Lake. McAdoo informed Wilson that he was not certain what, if anything, could be done to stem the tide of the current shark plague. The secretary insisted that Wilson at least mobilize the full scope of the U.S. Coast Guard, the National Marine Fisheries Service, and dispatch a federal agent to the shore to organize a battle against the man-eaters. Wilson had no other choice but to concur, and he also scheduled a full meeting of the Cabinet members to discuss the crisis. In an election year, when a natural disaster could inspire economic disaster, speedy and decisive action had to be taken.

The Vansant and Bruder attacks had already cost the Jersey Shore proprietors thousands of dollars in income, and governmental officials were even concerned about fallout for the upcoming season. The ripple effect could cause undue catastrophe. For instance, if a family felt that New Jersey was unsafe for bathing, they would simply make seasonal reservations with another hotel in Massachusetts for the next summer. Such a circumstance would have an effect on food ordering, equipment, employment, transportation, etc. Ultimately, the outcome would have a direct bearing on the careers of those in charge of correcting the crisis, including President Wilson. Eventually, the House of Representatives would appropriate $5,000 to search for and eradicate the New Jersey shark threat. In those days, five grand was a lot of money, as much as half the total appropriated for the establishment of the entire Coast Guard in 1848.

Newspaper cartoons now portrayed Wilson's chances for reelection in November, using the shark fin as the symbol for his potential for loss. The black fin labeled "defeat" was shown slicing through shark-infested northeast regions. Other political cartoons of the day showed lawyers, represented by sharks heading toward a beleaguered sailboat, embossed with "Union Bank." At the stern of the bank boat, a chewed and legless victim dangled over the gunnel depicting "deposits."

On the morning of July 13, Commodore Fisher and his family departed Minneapolis for New Jersey. Besides being devastated by the news, they were appreciably dissatisfied with the vague telegram regarding their dear Stanley's demise. What type of freak accident could have struck a tailor, they wondered? Was he injured by a carriage, a motor car, or a street car (trolley)? At the station stop in Chicago, Commodore Fisher was staggered when he spotted the front page of the *Chicago Tribune,* which described the horrifying circumstances behind Stanley's untimely death. The headline read: SHARK KILLS BOY, MAN IN CREEK ON NEW JERSEY COAST, SECOND LAD BADLY BITTEN. The Massachusetts–born steamer captain simply could not understand it. After all the time he spent at sea, he had never seen a shark harm a man, yet in the narrow bayou in the small town of Matawan, his son had been mauled and killed.

Meanwhile, at the American Museum of Natural History, Dr. Lucas assembled an emergency meeting with Dr. Nichols, the fishes expert, and John Murphy, of the Brooklyn Museum. The gentlemen were having the most frustrating and shocking month of their academic careers, but they were each men of science, and they realized that they must assimilate and make sense of the new developments. It would be a tough task for these gentlemen, since the news was not only startling, it was contrary to everything they formerly believed and publicly professed.

Lucas had been increasingly and directly confronted with transmissions from the beleaguered Commissioner of the National Marine Fisheries Service. He realized that Nichols had the deepest educational background when it came to sharks. For the moment, therefore, Lucas and Murphy were content (if not compelled) to defer to Nichols and ask for his assessment of the popular theories and scientific possibilities behind the attacks. Nichols began by stating, "People are saying that the ship sinkings and sailor deaths in the North Sea are creating a shark craving for human flesh. . . . Some are saying that the forceful naval bombings are driving dangerous European sharks across the Atlantic. Some are even speaking of the U-boats as a terror-driven cause."* Nichols appeared unconvinced of the sensational theories but did comment that he believed that a Pacific phenomenon called *El Niño* could have shifted the warm Gulf Stream a bit closer to shore. Murphy was also not silent on theories of his own and mentioned that such hunger may be attributed to an indirect U-boat effect, whereby the submarine warfare had diminished ship crossings and had therefore limited the usual staple of dumped refuse.* Near the end of the meeting, Lucas assigned Nichols to a fact-finding trip to Matawan to gather firsthand, eyewitness information. Before they adjourned, Lucas asked Dr. Nichols for his honest opinion of the developments. Nichols looked out the window at the trolley-packed street below and said, "I believe that this shark is moving north and attacking people on its journey. I believe it is either a white shark or a tiger shark that has strayed thousands of miles from its natural environment."† Nichols's hope was that the East Coast shark hunt, the largest scale animal hunt in history, would snare the man-eater.

Out on the street below, forty-five-year-old Michael Schleisser was carrying a stuffed warthog and circus posters to his apartment on East 32nd Street. He was accompanied by his long-time friend John Murphy (no relation to Brooklyn Museum Robert Murphy or the Cliffwood aunt of Joseph Dunn). Schleisser was the chief animal trainer for Barnum & Bailey Circus and one of the top taxidermists for several national museums. His work was displayed in the Smithsonian Institute, and in museums in New York, Philadelphia, and Boston. As he walked past the newsstand, Schleisser couldn't help but notice the *New York Times* top heading: WOULD-BE RESCUER AND BOY KILLED BY SHARK NEAR NEW YORK: ANOTHER LAD MAIMED. TOWN SCORNS TWO WARNINGS, MONSTER FISH TRAVELS 11 MILES UP SMALL CREEK. Once at his apartment,

---

New York Times, July 13, 1916.
New York American, July 14, 1916.
†*Brooklyn Museum Quarterly,* October 1916.

Schleisser commented to Murphy that it would be quite a sight to observe the shark-hunting frenzy near the creek in Matawan. Murphy objected to the idea, first citing the oppressive heat wave, but Schleisser finally convinced Murphy to join him on what he described as an invigorating and refreshing plan. The men arranged to depart from South Amboy, New Jersey, not very far from New York. There, they would borrow a friend's eight-foot motorboat and launch for a few hours on the Raritan Bay.

At the creek bank at Matawan, men hung from the few bridges along the waterway huge legs of lamb and sides of beef on large hooks. They debated which form of bait was most tasty to the sea wolves and spoke of the scores of newspaper reporters who had been checking into the Matawan House Hotel. The gentlemen also heard that film crews would be arriving to take motion pictures of the underwater blasting.

At the Matawan House Hotel on Main Street, which was already bustling with reporters, the hotel's enterprising manager, James Fury, realized that, as the town's main hotel, his facility would become the media's headquarters. He hastily hung new potted plants to adorn the entranceway and installed three new electric light poles.

A demanding reporter rang the desk bell and inquired about locating any attractive town ladies for a photographic pose at the creek. The clerk swiftly volunteered the services of his young cousins. At about 8:00 P.M., in the midst of echoing blasts of dynamite, the clerk peered over the desk to find the tall, distinguished Dr. John Nichols standing on the other side. The clerk said, "I bet you're a reporter."

The unamused Dr. Nichols replied, "No, I'm an ichthyologist."

"Oh," the clerk responded, "I thought you were here about them shark attacks."* Little did the small-town lobby clerk realize, but at the time of his visit, Dr. Nichols was working for the American Museum, the U.S. Department of Commerce, the Treasury Department, the

*Matawan Journal, July 13, 1916.

Federal Bureau of Fisheries, the State Bureau of Fisheries, the Coast Guard, and the president.

On Friday, July 14, from his office in Washington, D.C., H. F. Moore, Acting Commissioner of Fisheries, penned an urgent letter to Dr. Lucas. In the correspondence, Moore wrote:

> I have observed the newspaper press that Dr. Nichols has been sent to Matawan to make inquiries concerning the shark or sharks which have recently attacked persons at that place. We have had a number of inquiries as to whether it was our purpose to make such an investigation of this kind and have replied to the effect that it was not, as Dr. Nichols, who is already on the grounds, would be able to gather all the facts that it will be possible for us to get. Will you kindly arrange that Dr. Nichols should send us as promptly as possible anything that he may learn. I assume, of course, that there is no question but that the recent fatalities were due to sharks.

Along the Jersey Coast and up and down the eastern seaboard, "shark" was now a household word. New Jersey governor James Fielder had virtually mandated that steel-wire mesh be placed at all popular bathing areas. Asbury Park, Wilson's announced summer capital location, was now starting to become the town to where all roads led. Asbury Park was considered to be an innovatively constructed resort since its inception in the 1870s, and in 1916 it was even being eyed as a summit location for a settlement in the distant Mexican conflict. It was no surprise, then, that Treasury Secretary McAdoo scheduled a structured conference at Asbury Park's convention hall to discuss solutions to the crisis.

While private citizens were still out collecting bounty for sharks, sensible U.S. officials quickly realized that an effort at mass extermination of the sharks was futile, if not unnecessarily inhumane. Indeed, the futility of the endeavor had already received grand print coverage. Headlines read: SHARKS ARE MASTER OF THE COAST. Government officials and Coast Guard crew knew that protection was better than mass killing.

In attendance at McAdoo's conference in Asbury Park, which took place on Saturday, July 15, were commander of the U.S. Coast Guard, Captain Commandant G. L. Carden, District Superintendant Cole, and most of New Jersey's Coast Guard station keepers. The immediate concern for all officials involved was the success of the fence netting. They realized that if a poorly constructed screen were to fail, the calamity would destroy all confidence in that method of protection. Such an event would not only tarnish the image of a potentially effective form of intervention, but, just as important, it would cause economic disaster along the coast. The harsh reality was, if the nets failed, there wasn't much more to provide.

On the day before the conference at Asbury, Commander Carden was on the cutter *Mohawk* when he received "telephonic instructions" from "headquarters" and was to personally ascertain what protection, if any, could be accorded the public against further shark attacks. Carden wrote:

> In view of the general alarm occasioned by the loss of human life from sharks, local authorities and the public generally were *demanding* protection and the Government was obligated to provide it. Many wholly impracticable calls for assistance were being made, not the least of which were demands for patrols on the part of the Coast Guard vessels.

In fact, it was Carden himself, not McAdoo, who inspired the conference at Asbury Park to discuss shark protection. He directed each of the station keepers to inspect the local bathing establishments and conduct interviews with the managers. Carden, however, would *personally* interview "the local authorities at the more *important* points. . . ." At one of these "important points," Frank Smith, secretary of the Ocean Grove Camp Meeting Association, reported that "a very general alarm existed among the frequenters of the New Jersey bathing resorts, and that . . . he was at a loss to know just how to proceed." Smith also stressed that the question of expense was "merely relative and that the real consideration was having to do with the best arrangements obtainable." From Ocean Grove, Carden proceeded to Allenhurst and the vicinity. At the foot of Deal Lake near Loch Arbour and Allenhurst, just north of Asbury Park, the Coast Guard station keeper, William Van Brunt, was assigned to inspect the local nets and send correspondence to District Superintendent Cole and Coast Guard Commandant Carden. It is important and interesting to note that it was only one year prior that these Coast Guard stations along the beach were called U.S. Life-Saving stations. Their sole duty at that time was to assist people and cargo as it related to shipwrecks. In 1915, the Revenue Cutter Service merged with the Life-Saving Service to form the Coast Guard. It was still very much a gray area as to whether it was even appropriate for these station keepers to take any action in assisting bathers against the threat of shark attack.

Despite the confusion relating to protocol, Keeper Van Brunt had a strong sense of duty and did respond to his superiors' request. His letter reported on the inspection of the beaches from South Elberon to Asbury Park. These were the beaches with Gatsbyesque homes built on picturesque elevations just west of the sea. Such locations were not public resorts, but the more private homes of the very demanding upper crust. In 1920, F. Scott Fitzgerald wrote of this society, in *This Side of Paradise,* "they left the car at Asbury and street-car'd to Allenhurst, where they investigated the crowded pavilions for beauty." This region was made up of the individuals most likely to cause problems for the politicians if anything went wrong. During the fall of 2000, it was these very shoreline communities north of Asbury Park that delivered the greatest amount of scrutiny to a multimillion dollar federal sand-replenishment project. In 1916, at Deal, Van Brunt noted:

Poles all driven, ready to string wire, as soon as the surf permits. At Loch Arbour they are waiting for the sea to moderate but all material is on the grounds for work. At Allenhurst, they are in favor of some substantial structure for this protection and the borough council will discuss it this A.M. At Asbury Park, all bathing grounds around Asbury Avenue and Fourth Avenue sections were fenced in. The Seventh Avenue grounds were under construction.

Carden concluded that a heavy wire mesh, "not less than No. 9 gauge steel," would be adequate for the netting. Carden did not believe that it would be practical to completely enclose entire bathing stretches. He envisioned the enclosed areas to function more as pools, not more than one and one-half acres "of sufficient strength to insure absolute immunity from ingress on the part of dangerous deep sea fish." The commander also made note that the lessees of Asbury's beach facilities, Messrs. Mitchell and Fry (who had a three-year lease), were threatened with heavy financial loss if something was not done to solve the problem. Those entrepreneurs had already driven poles and had the idea of stretching heavy screen around the bathing concessions. Carden received word from Beach Commissioner George D. Morrow in Allenhurst, a private town that had traditionally put much of its energy into the facilities for summer enjoyment. Based on Morrow's comments, Carden wrote:

> Mr. Morrow intimated that no expense would be spared at Allenhurst to install a thoroughly efficient screening for bathers, and I was led to believe that some form of piling might be put in, either wooden or concrete, which would make of the beach over an area of one to one and one-half acres, a veritable swimming pool with interstices only sufficiently large to admit free ingress and egress of sea water, but not sufficiently large to permit of any large fish gaining access to the pool.

Today, I'm happy to report that most of the communities referred to on this last northern leg of the protection effort are tremendously well-preserved from an architectural standpoint. The communities of Allenhurst and Ocean Grove are now historic districts and Loch Arbour and Spring Lake are also eligible for such protection of their integrity. Asbury Park has preserved some of its classic waterfront structures, and it is in the thick of a promising restoration and revitalization. Essentially anyone can come down to the northern Jersey shore today and enjoy the grand Victorian architecture of that 1916 epic.

Commander Carden ordered all station keepers to avoid giving any information to the press. The superintendent actually referred to the precautions and mission of the Coast Guard as "top secret." Carden also believed that the sharks were only attacking in deep channels or near inlets. The shark at Matawan, Carden mentioned, had gained access by "penetrating inlets through deep channel approaches." At Beach Haven, he stated, Charles Vansant had gone swimming between the two sandbars that run parallel to shore. In that slough, between the shoals, was a deep channel. The stretches of beach area north of Barnegat, New Jersey, were said to be at greater risk of attack because the water becomes deeper quicker because of the steeper bottom slope. Commander Carden additionally mentioned that Atlantic City had a perfect opportunity to satisfy concerns because its beaches were already patrolled by "one hundred life guardsmen." He also felt that the practice at Atlantic City of routinely swimming at the inlets was unnecessarily dangerous.

Behind the scenes, specifics about fencing construction were now being conveyed. The metal posts recommended for the netting were to be not less than five inches in diameter and driven to a depth of at least eight feet. They were to be anchored by "claw foot anchors" and the use of chain at the lower edge of the netting was also suggested to keep the net on the bottom of the sand.

It's interesting to confirm the accuracy of unusual witness statements, as one mulls through the voluminous old records. In regard to Dr. Robert Patterson, at Spring Lake, I once thought he was misquoting when he referred to chains in the surf. As we are told, he said, "they put chains in the bathing area." Patterson even claimed to have seen a small shark caught up in the chain. Dr. Patterson, however, turned out to be right on the money. On July 13, the Spring Lake councilmen were visited by a representative from the American Chain Company of Bridgeport, Connecticut. Some officials even contemplated putting a pulley system on the fence poles to lift it during storms, similar to a sailboat mast. This pulley system was in use already by the old fish-pound netters. From a practical point of view, it was understood that the wire netting would not remain up all season without intermittent repair. In a lengthy letter following the Asbury Park conference, Commander Carden wrote to the director of U.S. Fisheries and justified the expenditure of funds for the purpose of protection and stated, "This is a legitimate overhead expense and is part of a burden which will have to be borne if the public is to be safeguarded." The commander's letter ended with this warning: "Bathers who venture outside the pools or beyond the lines of surf should be warned by the local authorities that they do so at their own peril." Superintendent Cole's comments to the Fisheries Bureau were more direct than Carden's. He simply said that the only way to protect against an attack in an unenclosed bathing area was *not* to go in. He added, "*I* wouldn't go in."

Although the Treasury Department, the Coast Guard, and the Bureau of Fisheries were in charge of the "shark problem," the Navy Department was also receiving earnest, if not helpful, suggestions from citizens. Max Silberman of Baldwin Street, New York, wrote to the secretary of the Navy and suggested: "Take a dummy and stuff it with the strongest poison possible and dress it with any garb that would be easy to bite off." Silberman also suggested attaching the dummies to steel or chain connections to immovable pillars. Silberman ended his letter by saying, "This plan may not suit, but there is only one thing in this world, than to 'try.' "

Cole also had to make a decision on whether to get the Coast Guard surfmen back from their summer breaks. The Coast Guard surfmen at the shoreline stations, other than the keepers, were traditionally off until August because the weather was routinely calm in the summer (i.e., there was no real shipwreck threat). Cole contemplated having them return from their leave but figured that this was such a new and uncontrollable problem, he wouldn't accomplish anything by it. Cole may have even anticipated the policy controversy that would arise if he made such a move. One keeper wrote an emphatic letter on the subject to Congressman Burke, stating that the concern over unsafe bathing conditions in his region "would not and could not be remedied by Coast Guard crews." The angered Coast Guard station keeper stated that Congress established the Coast Guard for "the preservation of life and property from shipwrecked vessels," not the protection of bathers. A keeper from Manasquan, New Jersey, did, however, make it his business to attend a borough council meeting to impress the importance of allocating monies for the protective netting. Most of these keepers, we should remember, were persuasive old salts themselves and came up with very *convincing* arguments when necessary. In this case, the Manasquan keeper, Andrew Longstreet, mentioned: "If the work were not done [on the fence netting], and anything should happen [e.g. an attack], the glaring headlines of newspapers would make them [the council members] feel small enough to get under a thimble." Longstreet's correspondence to Superintendant Cole explaining the encounter with the Manasquan commissioners added, "After having been advised, it went a good ways in making them cough up." General Coast Guard correspondence from that period also told of beach proprietors that simply refused to enclose their bathing areas because the number of bathers did not warrant the expense of the task.

Speaking of peril, it was found that a person's general accident insurance policy would cover him in case he was attacked or maimed. No one seemed to think that a company would contest a claim after an attack. An insurance company spokesman additionally felt that action could be brought against the proprietors of the bathing beach where the accident occurred. The head of a Workmen's Compensation Board claimed that only persons employed at the beach could seek such compensation. One insurance officer made this comforting confirmation: "There is nothing prohibiting bathing in the ordinary accident policy, and anything happening in the water would have to be classified as an accident."

Beyond the jurisdiction of the federal government and Coast Guard, shark hunts were still well under way. The Paterson Chamber of Commerce in northern New Jersey (nowhere near a beach) offered bounty money for sharks captured at rates proportional to the shark's weight. Governor Fielder delivered suggested bounty rates to Seagirt on July 14. The rates were $2 for a 110-pound shark, $5 for a 200-pound shark, $100 for a 300-pounder, $25 for a 400-pound shark, and $50 for a 500-pound shark. The $100 tag on the 300-pound shark was either a misprint (meant to be $10) or those setting the bounty must have believed that the accurate witness testimony from the creek revealed a "300-pound shark." (Such an assumption will be interesting later.) They could have also figured that 300 pounds was a good man-eating size. Wildwood, New Jersey, offered the generous reward of $1,000 for any proven man-eater, but to meet the man-eating criteria the shark had to contain human remains when dissected. By the way, the offered bounty from *Amity* for the killer in the 1975 motion picture *Jaws* was $3,000.

Even though much of the wire netting placed up over the Matawan Creek was nothing more than a show of emotion, an International News Wire service dispatch sounded quite dramatic about its failure. On the evening of Thursday, July 13, it told of the man-eater of the creek cleverly eluding the shark posse that had placed the stout wire fencing across the creek mouth. The fencing barracade was supposedly "strong enough to repel an attack by any shark," but on that night "a giant shark bucked the line like one would do in a football rush."

On Friday morning, July 14, Schleisser and Murphy departed in an eight-foot motorboat from South Amboy. Before they left the dock, Schleisser accidentally broke one of the only two wooden boat oars but decided to bring the broken oar handle along for the ride. As they headed out toward Raritan Bay, the men had no idea that this short boat cruise would turn out to be the ride of their lives.

During the same sunrise that Schleisser and Murphy departed from South Amboy, at 5:30 A.M., train conductor Harry Van Cleaf took his customary walk to work along the dirt path that follows the Matawan Creek from his house. For Van Cleaf, however, this walk would be far from customary and one he would never forget or discuss. The overnight deluge had made the ground soggy under his work boots. At the train trestle portion of the creek, about 150 feet west of the Wyckoff dock, Van Cleaf stopped abruptly and strained to visualize the brown water through the dripping leaves. There in a shallow pocket of water under the overpass, he recognized the pale figure of Lester Stillwell floating face down in the muddy creek. The corpse was noticeably tattered and nude.

At the home of Mayor Henderson, Mrs. Henderson provided her husband with a second cup of black coffee as he waded through a stack of articles and telegrams. In the weeks to follow, the mayor would also receive stacks of letters from around the country. Every item offered "expert advise" on how the town could rid the creek of the man-eater. Annette Kellerman, Australia's leading woman swimmer, wrote that "the shark is at heart an arrant coward and will flee at the slightest disturbance, that is if he is well fed."

A guest at the American Hotel in St. Louis, Missouri, could not resist getting essential information to Mayor Henderson. The man wrote:

> Dear Sir,
> Many years ago I had a friend in the island of Barbados whose hobby was shark hunting. Whenever he could secure a dead horse or mule, he would mutilate the carcass and then have four negros tow it out into the bay. Invariably, sharks would attack the carcass and he would shoot them with a rifle from the

stern of a boat. I know the plan worked in Barbados and I think it would be worth a trial as it is quite inexpensive.

Yours very truly,

T. Allen Lourie.

A woman from Denver, Colorado, wrote: "My father, a native of the West Indies, suggests the following: Construct a raft capable of floating a dead horse partly submerged and anchor it in the vicinity of sharks. Then, locate your motorboats and dories within an easy range. This floating object unfailingly draws the man-eaters and no difficulty is experienced."

A man from Bradford, Massachusetts, couldn't help but tell Mayor Henderson of his discovery sixty-seven years earlier. The man found the vital addition of metal wire fishing leader as an aid in catching sharks. The letter states: "On a trip in 1849, in the Southern Atlantic, one big shark in particular had followed us for several days. A Cape Cod boy of fifteen years caught the shark by wiring his line [metal wire as opposed to fishing line] from the hook. If your people want to catch your sharks, you must wire your lines or they will be bitten off every time!"

An innovative suggestion came from Philadelphia, where a man, after learning of the attacks on Stillwell and Dunn, assumed that a shark(s) had acquired a particular taste for boys. The man wrote: "Would it not be a good idea to make a dummy out of flesh colored clothing and sawdust about the size of the boy attacked and attach wires from a battery that would explode dynamite concealed in the legs of the dummy?"

Alex Murdoch Jr., of Landsdowne, Pennsylvania, sent the Matawan officials a suggestion for another explosive device. Murdoch's detailed sketch/diagram revealed a ball of meat attached to strong piano wire hanging from a spring-loaded float. The instructions read: "When the shark pulls the bait he pulls up the spring and makes contact with the electric circuit which sets off the dynamite."

The Cyclone Fence Company of New York heard that fencing was enlisted in the defense against the man-eaters and sent a telegram to Henderson that read: "Shark horrors can be promptly eliminated! Wire us for a representative at our expense."

Sam Groves of Wilmett, Illinois, wired another telegram to Henderson that just about knocked the mayor off his stool. This telegram was a suggestion presented in question form, like the one about the sawdust dummy in the shape of Lester Stillwell. Groves wrote: "Why don't you have the U.S. Government round up all available submarines to hunt that murdering shark? Use beef bait if necessary."

Henderson even received a fairly practical suggestion from John S. Clarke, vice president of the Autocar Company of Ardmore, Pennsylvania. Clarke pointed out that the system of using motorboats and guns was "altogether too uncertain and the sharks are quickly scared a distance from the engine commotion anyway." He suggested a series of white-painted kegs attached to baited lines that he utilized to clear shark-infested Tarpon grounds in the Florida Keys.

Although Mayor Henderson was knee deep in correspondence about the shark catastrophe, he was not the only Matawan official to receive "helpful" suggestions. Frank Mulsoff, town barber and Matawan police chief, received an interesting if not bizarre note from one enterprising individual. The letter to Mulsoff, written in pencil, was written by a man who would not divulge his foolproof plan unless he was guaranteed a private meeting with town officials. The man writing to Chief Mulsoff began by saying: "Dear Sir: I have been reading of those man-eating sharks and have a very easy way of catching them at very little expense. So if you are interested, answer by return mail at my earliest convenience to explain my plan. Very truly, Herbert Van Fleet." At the bottom of Mr. Van Fleet's bold letter, he wrote, in a less than straight line: "P.S. Please excuse pencil."

The suggestions making their way to government officials often had a competitive tone. Secretary McAdoo received a letter from Edwin Hewitt (of the Hewitt Candy Company) of Denver, Colorado, who apparently had extensive experience with shark hunters from Texas. Hewitt wrote: "It may seem strange that a man 'one mile above sea level' would have the crust to offer suggestions in connection with the 'war upon tiger sharks,' but if the government desires to wage a war of extermination upon these ferocious beasts of the sea, go to Port Aransas 'Tarpon' Texas." The letter was attached to five photographs of prize fish, and the authors emphasized that this was not a publicity stunt but the "only way" to accomplish the objective with minimal time, expense, and risk. It's interesting to note that Port Aransas was the sight of a vicious shark attack in April 1987.

The most impressive credentials to appear before any Matawan officials (or any officials for that matter) were from a charter boat captain from Miami, Florida, named Charles H. Thompson. Captain Thompson would not only have made a perfect 1916 version of the legendary Montauk shark hunter Frank Mundus, but he would have even been able to teach *Jaws*'s Captain Quint (Robert Shaw) a thing or two about self-promotion. Thompson's letterhead titled him: Capturer and Owner, Monster of the Deep. In no uncertain terms, Thompson informed the governing body of Matawan that he was the man for the job. Thompson wrote: "After reading the account of the terrible shark tragedy, it occurred to me that I could be of assistance to you and do a great deal of good to the community by killing out the monsters for you as I have made a lifetime study and business of capturing and killing sharks. I know I can rid your coast of the dangerous sharks if your city will cooperate with me. I can furnish you with all the references you wish as to my ability in that line."

After making it to the last letter it must have been a relief to Henderson to know that he'd at least be meeting with a real scientist from New York that morning near the creek. John Nichols was on his way to Matawan. Perhaps he'd be the voice of reason.

At the quiet Stillwell home, Sarah Stillwell had just come out to the front door with a drink of water for William, who was still chopping wood. She had a cup of tea waiting for her at the porch table. She was joined by their remaining son, John. Before she reached the steps, she spotted someone walking toward the house at the end of the dirt road. Her face went blank, and she dropped the glass of water. William looked at her, then down the road at the object of her attention. Stillwell sternly and calmly ordered John back inside. William took a final swing of the axe to rest it in the chopping block. Along the dirt path in front of the house, stone-faced Harry Van Cleaf carried a blanket-covered bundle. The lights turned black for Sarah Stillwell, and she collapsed into her rocking chair. She attempted to get up but immediately fainted again. William Stillwell took the mangled remains of Lester from Harry. No words were exchanged.

The late Mildred Van Cleaf Fisher, the daughter of train engineer Harry Van Cleaf, was gracious enough to fill me in on her father's response to finding the body of Stillwell. Mildred vividly recalled her father's morning routine, but as for the discovery of Lester, it seemed as if it was an intentionally surpressed subject. Apparently, her dad would never retell the story of what he had discovered on that damp morning near the trestle. "So every morning," she began, "my father used to take a walk. He used to go down one street and walk along the water's edge and then cross and go up to Matawan railroad station. And then one morning that's what he saw [Stillwell's floating corpse], but what he saw I don't know. He did run up to the station when he made the find and told them about it. We [her family] didn't even know he had seen the body until it came out in the newspapers. It may be that the sight he saw," she said with emotion, "he just didn't want to discuss it. He never brought it up." Mildred's roots and legacy in Matawan are still vibrant, and I believe that one of her grandsons is the young mayor of Matawan at this very time.

At the Matawan Creek, now teeming with reporters and townspeople two days after the attacks, manually cranked motion-picture cameras lined the sloped shoreline. Geyserlike blasts of water were captured on film by the North Jersey and New York film crews. Some photographers got members of the fairer sex to pose and gingerly point rifles in the direction of the water. By midmorning, one New York newspaper representative arrived at the Wyckoff dock with a large motorboat dragging lamb-baited grappling hooks off the stern and advertising a shark-hunting expedition. Later that morning, seven baited hooks hung from the Matawan trolley drawbridge and twisted; ineffectual chicken wire fencing was visible up and down the creek at low tide.

The *San Francisco Chronicle* of Friday, July 14, displayed a huge front-page headline that summarized the current state of affairs in New Jersey: EAST COAST BEGINS WAR ON RAVENOUS MAN-EATERS . . . BATHING ALMOST DISCONTINUED AT FAMOUS RESORTS EXCEPT WHERE WIRE FENCES ARE USED FOR PROTECTION.

The late John Applegate, former Matawan mayor and a town kid in 1916, recalled the influence of the bizarre creek-side action on barroom conversation in later years:

> I can remember how jokes would evolve, even over very serious things, especially in the barroom in those days. A woman was never allowed in a bar. Never! At least not the one in Matawan anyway. And the men would get talkin' and they'd be telling about some damn fool in a rowboat towing a piece of steak behind him. Or about people throwing dynamite in the crik and killing a lot of good fish and all that.

On a bluff situated above the unorthodox creek-side activities, Dr. Nichols was speechless as he stood with Mayor Henderson and observed the unscientific media frenzy. Nichols now represented the hope and heroism of science. He was the real-life character who would be mimicked in the form of baffled and brave professors in 1950s' science-fiction thrillers. The mayor informed Nichols that the rainstorm of the evening before helped nudge the corpse of Lester Stillwell to the surface of the creek at dawn. Nichols was interested in examining the body of the boy, but Henderson told him that it had already been moved to the Arrow-smith Brothers Funeral Home to undergo embalming preparation. The medical examiner did see the body, and Henderson was holding the autopsy report. The report describes the wounds as delineated by the coroner who came up from Red Bank. It read:

> The boy's left ankle was chewed off, his left thigh was mangled from hip to knee, his left abdominal region was ruptured and the intestines herniated and torn open. The intestines were nearly all torn out. The right hip, right chest muscle, left shoulder, as well as several fleshy areas of the body were all eaten away and the flesh between his right hip and thigh were mangled. His face was untouched.

The formal death certificate was more to the point. The cause of death was recorded as "Bitten by shark while bathing in Matawan Creek," and the contributory or secondary cause as "Hemorrhage & Shock." Lester's dad wrote in the information for the personal section of the certificate and recorded Lester's occupation to be "School" rather than designate him as a summer factory apprentice.

Nichols's presence in Matawan was not merely to gain information on the circumstances of the attack. He also wanted to increase his odds of determining the reason behind the awful string of tragedies. He and other scientists throughout the United States and the world had never seen something so dreadful and inexplicable. This was all without documented precedent. With all the theories circulating about assorted sea creatures perpetrating the events, Nichols wanted to identify this phantom marauder firsthand. To the scientists of the era, this was uncharted territory. Nichols wasn't quite sure what he would find, but if it was something unusual or heretofore undiscovered, he wanted to be the one to find it.

After the first two attacks, the public and the scientific community had a fairly certain opinion that the attacks were being perpetrated by a shark. For whatever the reason, however, the American Museum group, or at least Dr. Lucas, was still being noncommittal about implicating any one type of creature. Now, the witnesses at the creek all but confirmed that it was a shark responsible for the horrors. Although the creek water was far from translucent, the opportunity for viewing the terrible action, at least in the case of Fisher, was incredibly conducive to clear observation. The crowd was on the nearby dock anxiously watching to see if the men were going to find young Lester. During the attack on Fisher, he and the shark were just a few feet in front of the onlookers. I would venture to say that never in history has a shark attack been so publicly visible and close in its proximity to vigilant spectators. With a forum so stagelike, and the events so bizarre, I doubt it could ever happen again.

When Nichols awoke that morning, the front-page headline in the *Newark Star-Eagle* must have been quite an eye-opener. The *Star-Eagle* (now the *Star-Ledger*) was, and still is, one of the most popular and widely circulated New Jersey newspapers. The paper on that Friday reported: SHARK KILLED AT KEYPORT; HUMAN BODY INSIDE. SHARK TERROR SLAIN NEAR SCENE OF TRAGEDY. The article described the capture of a large shark, which, when cut open, revealed the body of Lester Stillwell inside. The grisly rumor started when a false report circulated through Keyport and Keansburg stating that an eleven-foot shark weighing three hundred pounds had been captured at the mouth of Matawan Creek. Captain Collins, a fisherman from Keyport, telephoned the news to W. L. Martin, the "cooperating observer" of the U.S. Weather Bureau at Long Branch. The report also included the news that the boy's body was found in the shark. Martin telephoned the story to the New York office of the weather service, and the story became the buzz around the city.

For a moment, just imagine being the editor of one of these hungry newspapers. In 1916, newspapers represented the king of media formats and the ultimate informational tool to the masses. On July 12, an unprecedented and gruesome tragedy occurred in which a boy was killed and a carnivorous sea wolf was responsible. The youngster's remains were unaccounted for and he was presumed to have been consumed by the monster. Immediately following the slayings, the bounty hunt was bigger than the search for Pancho Villa, John Dillinger, or the Lindbergh kidnapper. Any copy editor would be chomping at the bit to run a story about poor Lester's corpse being dissected out of a toothy terror, even if questionable sources conveyed the information. And so it came to pass; the *Star-Eagle* received the story that Lester's body dropped out of the dissected carcass of a man-eater. The man-eater was caught in an area that was buzzing with shark talk. There were literally hundreds of sharks being caught, but on this day, there were many forces pushing for a guilty party, and in the case of the Keyport shark, there was as much authentic guilt surrounding its publicized capture as there was with the initial large tiger shark captured in *Jaws*.

When Harry Van Cleaf presented the Stillwells with Lester's soggy, bloated remains, that was proof enough that the story by the *Star-Eagle* was far from accurate. The headline the following day made no dispute about the facts. It read: BODY FOUND; SHARK HUNT IS RENEWED: MUTILATED CORPSE OF LESTER STILLWELL DISCOVERED NEAR SCENE OF TRAGEDY.

At 10:00 A.M. on Friday, the somber Fisher family arrived at the Matawan station. Their neighbors, the Smiths, and the Fisher's longtime housekeeper, Tillie Holmes, were there to console them and update them on the arrangements for Stanley's burial. Fisher's sister, Augusta, had made the long trip all the way back to New Jersey to pay respects to her cherished brother. Commodore Fisher and the Stillwells decided that the funeral services would be private and the ceremonies would be combined. They also determined that the caskets would be closed for most of the services, but Stanley's would be opened for a short time at the end of the ceremony.

On the morning of Saturday, July 15, organ music bellowed from the private church services at the First Methodist Church. Mary Anderson stood conspicuously alone. No longer was the choir supported by the strong deep voice of Stanley. No longer were Lester's young companions able to whisper to their skinny buddy about the appearance of their schoolteacher. Stanley Fisher's family wept, and Lester's mom held the forearm of her steady husband. Lester's buddies lined the front pew. Two coffins, one large and one small, occupied the center aisle. Some of the mourners spoke about how Stanley had just opened his shop last March; others were too distraught to say a word. In a newspaper announcement that would come later in the week, Stanley's family would present a commemorative poem to the *Matawan Journal*. The poem's heading read: "In sad but loving memory of our dear beloved son and brother, W. Stanley Fisher, who was taken from us so suddenly July 12, 1916." Signed, "His bereaved mother, father and sister." The poem was composed by Augusta and was read aloud at the ceremony. The verses are as follows:

> *It was hard to part from him,*
> *The one we loved so dear;*
> *The heart no greater pain could feel,*
> *No sorrow more severe*
> *What pain he bore we will never know—*
> *We did not see him die,*
> *We only know that he had gone,*
> *and never said good-bye.*

The debilitating heat did nothing to distract from the emotional church services conducted by Reverends Leon Chamberlain and B. C. Lippincott. Their remarks were of a personal nature and paid tribute to the sterling worth and generous spirit of both of the deceased. The mass ended with the pallbearer procession, which left hardly a dry eye in the house of worship. Red Burlew, Arthur Smith, and Asher Wooley were among Fisher's pallbearers. Lester's four pallbearers were, of course, his young friends from the creek. The coffins were placed on horse-drawn wagons and began a slow and steady course toward Rose

Hill Cemetery. When the wagons veered off Main Street to venture up Ravine Drive, a massive row of automobiles was revealed. Nearly the entire town and residents from neighboring Keyport and Cliffwood joined an automobile procession of mourners.

At a central mount at the cemetery, Stanley was laid to rest at the family's plot. At a ground-level spot near a curve in the internal cemetery path, Lester was one of the first to occupy his family's burial location. At Lester's graveside, two women sang "Surrender All" and "Safe in the Arms of Jesus." The Sunday papers reported MATAWAN MOURNS WITH FAMILIES OF SHARKS' VICTIMS: GREAT THRONGS ATTEND FUNERALS OF MAN AND BOY SLAIN BY SEA WOLVES.

The haggard Fisher family returned home that afternoon only to find a young man waiting to greet them. They knew the man to be an acquaintance of Stanley, Ralph Gorsline, but because of the heat and the intensity of the day's events, Commodore Fisher explained that they were in no condition to do more than acknowledge the man's expression of condolences. Commodore Fisher encouraged the man to return in the coming weeks if he desired. The man immediately respected their wishes, but as he walked down the front walkway, he explained that he actually wanted to speak with Mrs. Fisher. As he continued to back up, he mentioned that he worked for London & Lancashire Indemnity and briefly told about how Stanley fatefully accepted a $10,000 life insurance policy in exchange for the sale of a suit. Mrs. Fisher was beneficiary under the contract. The Fishers were awestruck and arranged a meeting for the following Tuesday. The sum of the principal was $7,500.

When the Fishers entered their modest living room, they were thinking nothing of the benefits that the bizarrely procured life insurance policy would provide. Their thoughts were of Stanley and nothing else. No amount of money would bring him through the door or quell the torment. When the last tears of the day had fallen, the family decided that dedicating the insurance policy funds toward anything other than Stanley would be a great injustice.

Stanley's voice would be missed from the Methodist Church choir, but his memory would remain. With the policy money secured, Stanley's family decided to purchase and design a magnificent stained-glass window for the front of the Methodist Church and, according to one of the family members, they chose such an object "so that the rays of the setting sun would forever bring Stanley's spirit to the minds of those present."

The immense window was an exquisite piece of workmanship. The window's scene depicted a landscape of trees and flowers in the foreground and the city of Bethlehem in the distance with its historic mountains. The inscription below the window conveyed the essence of Stanley's selfless sacrifice. It read: "In Loving Memory of W. Stanley Fisher" and underneath was a quote from the Gospel of John 15:13: "Greater love hath no man than this, that a man lay down his life for a friend." At the window's dedication in July 1918, Reverend B. C. Lippincott referred to Stanley Fisher's attempt to retrieve Lester as an act of courageous and spontaneous love, which added beauty to a world that was being ravaged by a horrible war [World War I].

In Matawan, life did not end with the shark attacks. Life went on, and the hardworking townsfolk did not look back except to commemorate those they lost. Stanley's sister returned to Minnesota, and her parents grew old and passed on in Matawan. Augusta Fisher Nichols never forgot her beloved Stanley or his link to the Methodist Church. Every few years, until the 1960s when she presumably passed on herself, she would send donations in Stanley's name. In 1966, at the fiftieth anniversary of the attacks, the Methodist Church presented a memorial service for Stanley. The reverend spoke of the Bethlehem window. He also spoke of Stanley Fisher as "one of the most well-known and admired men in our town. . . . Over the past fifty years," the minister said, "the rays of the setting sun have filtered through this window as the day's end came to Matawan. Yes, 'greater love hath no man.'. . . Stanley surely deserved all the honor and glory bestowed upon him by his Heavenly Father."

The Fisher window was damaged by a local bomb blast no more than three months after its installation. Time and fading sentimentality, however, took their toll as well. The First Methodist Church fell to demolition in the 1970s, and its stained-glass windows, including the Fisher window, were auctioned off to the highest bidder. The current whereabouts of the Fisher window are unknown, and it is my hope that its owner recognizes the great value behind its origin. The details of the Fisher window may have been lost to the winds of time if Walter Jones, a Matawan postal worker and church historian, had not presented them to me in 1989.

Some of the memorabilia and strange remnants of those chaos-filled few days still remain in scattered private collections throughout New Jersey. Clark C. Wolverton of Orange, New Jersey, a former Matawan schoolteacher who once rented a room from Mayor Henderson's late widow, has an extensive collection of interesting original documents. When Mrs. Henderson passed on, Mr. Wolverton was fortunate enough to inherit a chest filled with letters and telegrams from around the nation. The documents speak of concern for the terrible shark situation. The chest also contained one of the infamous shark "wanted" posters. That assortment of materials is a prized possession of the Wolverton family.

During the seventy-fifth anniversary of the Matawan attacks, I was invited to speak before the Matawan Historical Society and show my original comprehensive video documentary on the subject. I was happy to present the historical society with the original hand-drawn copy of Leroy Smith's 1916 town map. I was also given the opportunity to meet with the Vineyard family, who represented some of the only remaining local descendants of the Stillwell family. Mrs. Vineyard informed me that one of the traditional family warnings handed down to some of the more adventuresome family boys was, "If you don't come straight home after school, you'll end up in trouble like Uncle Lester." One of the older aunts of the extended family recalls visiting the Stillwell home as a young girl and seeing young Lester's portrait hanging on the left side of the doorway just as she entered the home. "My aunt Sarah and uncle Bill kept that picture of Lester up forever but never seemed to speak about him," she said. I felt compelled to present that gracious woman with a copy of the photograph I duplicated from the newspapers of 1916. It is that photograph which I assume was hanging in the living room of the Stillwell household. The woman also informed me that Lester's stoic dad remained serious and workmanlike for the remainder of his years. When William Stillwell was in a failing elderly state and his aging heart told him he was in his last moments on earth, "he simply walked into the living room, laid down in front of the fireplace, and ceremoniously placed his hand-shaving blade in his right hand and gently rested it on his chest, [almost as if to say] his work was done."

Belford, New Jersey shark. New York American, *July 15, 1916.*

CHAPTER **5**

When the Sea Tiger Attacks

I n perusing the July 1916 shark-related newspaper headlines, it takes but a few turns of the microfilm wheel to find the term "sea tiger" or "sea wolf" used to characterize the shark implicated in the gruesome attacks. Although by today's standards it is inaccurate, the term "sea tiger," while sensational, does, in many ways, seem quite appropriate. Even in today's world, where television documentaries and magazine print present a thorough portrayal of the behavior of apex predators, we are still mesmerized, enthralled, and terrified by the thought of a large beast seeking out man as prey.

The only seminotable "attack" in New Jersey in 1917 took place in late September when a lifeguard received a knee abrasion from an apparent shark. Other than that, New Jersey and the world did not see any notable mention of shark attacks until the sea tragedies of World War II, the South African attacks of 1957–58, and the New Jersey attacks of 1960 (all of which will be discussed in detail later on). Even then, however, the press did not cover these stories with the same intensity and passion as it did in 1916. Certainly, ancient accounts from seafarers, contents of fictional maritime literature and film, as well as our natural fear of any large predator have contributed to our mortal fear of sharks. I can't help but think, however, that the 1916 attacks represent the pinnacle of the modern psyche regarding sharks. The story of those events has woven its way through generations into our culture. Perhaps it is the renditions of the events and dormant utterances that spring forth from time to time in our presence, along with the remnants of the 1916 attacks in popular and scientific literature, that has transferred the seed of an eighty-five-year-old tragedy from one generation to the next. During the New Jersey attacks of 1960 and even at the time of a Florida bull shark attack of August 2000, the infamous 1916 attacks were referenced as a standard of shark dread. Such a seed of dread likely began to sprout and blossom in the mid-1970s. One cannot forget the beach scene in the motion picture *Jaws* when, after the second attack, the troubled Mayor Vaughn (played by Murray Hamilton) discusses the crisis with Chief Martin Brody (Roy Scheider) who recalls that "rogue" attacks did happen once before, down in New Jersey, "in 1916."

Although our inherent fear of sharks is all too real, the frequency of such attacks is far less common than one would think. Throughout the entire world, there are between fifty and a hundred attacks each year, resulting in only five to ten deaths. Far more people are killed by bee stings, snakebites, mosquitoes, or lightning, and, as many authors point out, you are far more likely to be killed in a car accident on the way to the beach than by a shark in the water. Statistically, you are thirty times more likely to be struck by lightning than to be attacked by a shark, and one thousand times more likely to drown. Dr. Eugenie Clark once put fear of shark attack into clear perspective by posing the question: "As you walk down the sidewalk while cars are passing along the street, are you preoccupied by the apprehension that a car will come up the curb and strike you?" When one considers the number of people who swim in the sea or are otherwise unknowingly exposed to free-swimming sharks, the chances of being attacked are likely one in several million, and the chances of becoming a shark-attack fatality are higher still.

Regardless of the comforting statistics, sharks still deserve their reputation as perfect predators and efficient killers. The shark's biting, swimming, and sensory apparatuses are close to unparalleled in the underwater world. We often don't even think of sharks as regular fish, and indeed, the shark doesn't contain the normal bony skeleton that most fishes do. In his 1982 *Oceans* article, "Sharks Don't Swim—They Fly," American naval scientist H. David Baldridge described shark swimming abilities in aeronautical terms. The shark's tremendously strong but flexible muscular, cartilaginous makeup creates the general shape of a bullet or a spear. Its dark armor of sandpaper-like hide, which adorns the well-recognized dorsal fin, the caudal (tail) fin, and winglike pectoral fins, finish a figure that can only be described as an emotionless, phantom, fighter plane, or as Peter Benchley recently described the great white, "like a torpedo made of stainless steel." The prolific Dr. Robert Murphy and the distinguished Dr. John Nichols wrote in October of 1916 (*after* the attacks):

> There is something peculiarly sinister in the shark's make-up. The sight of his dark, lean fin lazily cutting zig-zags in the surface of some quiet, sparkling summer sea, and then slipping out of sight not to appear again, suggests an evil spirit. His leering, chinless face, his great mouth with its rows of knife-like teeth, which he knows too well to use on the fisherman's gear; the relentless fury with which, when his last hour has come, he thrashes on deck and snaps at his enemies; his toughness, his brutal, nerveless vitality and insensibility to physical injury, fail to elicit the admiration one feels for the dashing, brilliant, destructive, gastronomic bluefish, tunny, or salmon.

The shark's powerful jaws exert a force of 42,000 pounds per square inch (an adult male human bite force is 150 pounds per square inch). The bite is complimented by multiple rows of replaceable teeth, and it has even been estimated that a shark may shed as many as thirty thousand teeth in a thirty-year lifetime. Indeed, shark's teeth are the number one most collected animal artifact.

The shark's ability to detect odors is likely its most acute sensory system and is employed to zero in on distant prey. The shark can detect specific odors up to one mile away and routinely can detect one part of blood mixed with 25 million parts of water. Sharks have even been known to be attracted to a diver's nosebleed, but research suggests that some sharks may prefer fish blood to mammalian blood. If deprived of food, the shark's acuity for odors can increase severalfold. Sharks are also attracted by other forms of odors such as feces, garbage, and vomit.

Besides its sense of smell, the shark utilizes a keen hearing system to detect even the slightest vibration from a hundred to a thousand feet away from an object. A fish in distress or a bather splashing are easily picked up by a shark's vibratory senses. Since all animals generate and give off electrical fields, it is no surprise that sharks possess unique organs, called the ampullae of Lorenzini, to enable them to locate prey by keying in on the prey's natural electric field. These uniquely sensitive electric-detecting organs are the very reason why electrically based shark repellant systems are being sold and perfected at this moment.

Closer to its quarry, the shark's vision comes into play. It can see as far away as fifty feet and is very sensitive to movement and contrasting colors. Sharks also have a special structure to amplify ambient light to make night hunting possible even in the dimmest of conditions. For sharks that are mainly daylight feeders, like the great white, the sight of swift movement or contrasting shades (like the soles of one's feet or tan lines), may be all that is needed to trigger an attack in the murky surf. Natives in specific shark-infested locations sometimes cover the soles of their feet before harvesting oysters and fish.

A feeding shark that is subjected to unfavorable surf conditions must make quick decisions to capture its traditional prey. Most people do swim close to shore, but it's no coincidence that attacks typically occur inshore of a sandbar or between sandbars, where sharks can become trapped at low tide. Sharks generally prefer deeper waters, only coming close to shore at night. This fact also correlates with the peak hours for attacks (between 2:00 and 4:00 P.M.), but one should also consider that this is the peak hour for human presence in the surf.

When considering the impact of water depth, distance from shore, water clarity, sea conditions, and location as they relate to shark attack, one should always remember that much or most of the influence points back to human traits rather than shark preferences. Since most people swim in shallow areas, for example, most attacks (62 percent) have taken place in less than five feet of water during calm seas (69 percent), on weekends (65 percent more than weekdays) when a person is swimming, wading, or diving, as opposed to board surfing or bodysurfing (8 percent). Similarly, over half (51 percent) of recorded attacks have taken place less than two hundred feet from the water's edge (with the greatest number being within fifty feet of shore), but this is also the area where most bathers are found. Considering the fact that far fewer swimmers venture beyond the two-hundred-foot mark, closer analysis does reveal a proportional increase in the number of bathers attacked when swimming between the two hundred feet and one mile location. Gender and age also reveal a definitive statistical impact where Caucasian males, ages fifteen to twenty-four years, account for the vast majority of attack victims with a male to female ratio of 13.5:1. Reasons for this discrepancy are not fully understood, but it has been suggested that males tend to be more active in the water and generally venture farther from shore. The ratio mentioned is also quite similar to that encountered for drowning deaths. Channels and areas where the water suddenly deepens are favorite spots for sharks, which use topography to stalk prey without being seen. Inlets, especially near harbors, docks, wharfs, jetties, bays, rivers, and river mouths, are also potential attack sites (18 percent) because they are often used by sharks for breeding grounds and also contain edible refuse.

As far back as 1933, Dr. Victor M. Coppleson, from Australia, and in the 1950s, Dr. David H. Davies, of South Africa, suggested that water temperature played a key role in determining the chances of a shark attack. The scientists recognized that approximately 80 percent of attacks took place in temperatures between sixty-eight and seventy degrees Farenheit and contended that the chance of shark attack below a water temperature of sixty-eight degrees was supposed to be low. Some twenty years later, Baldridge reevaluated this contention and, after studying beach population density as it related to climate and water temperature, determined that the statistics on water temperature relate more to people's comfort and physiologic ability to maintain body temperature than it did to shark behavior (i.e., there are more people in the water at these temperatures). In Florida, the shark population is actually greater in waters colder than sixty-eight degrees (sharks prefer waters below sixty-eight degrees) and sharks are known to inhabit and flourish in much colder regions. Hence, modern shark knowledge suggests that the sharks are always "there and capable of attacking," but people are only present at preferred times.

According to George Burgess, director of the International Shark Attack File (ISAF), there are three forms of unprovoked attacks. The most common attack display is the "hit and run" version, in which a victim rarely sees the attacker. This is the type of attack that most commonly occurs near shore, in murky surf, where a swimmer or a surfer is mistaken for a seal, sea turtle, sea lion, or a fish. Such an attack scenario and the fact that very few attacks involve more than one bite has inspired people like the ocean explorer Jean-Michel Cousteau to refer to sharks as "man-biters," not "man-eaters." Cousteau explains that sharks use their mouths and teeth to examine the prospective prey item and then release it when they discover it is not what they expected. Injuries sustained by "hit and run" victims are usually limited to small below-knee lacerations. According to some sources, 94 percent of all shark attacks are perpetrated by a single shark acting alone. In only one third of cases was the shark seen prior to the attack.

The "bump and bite" and "sneak" attacks, while much less common, result in more serious injuries and are responsible for most attack fatalities. Such occurrences usually take place in deeper waters and involve divers, swimmers, downed pilots, and shipwrecked sailors. In the "bump and bite" version, the shark circles and bumps the victim prior to the attack, which then involves several vicious bites. The "sneak attack" occurs without an initial warning by the shark and is probably the result of actual feeding or a defensive act by the shark against what it considers to be a threat.

Since we know sharks and people have been sharing the planet for a long time, we would expect to have a plethora of written accounts relating to early attacks. Unfortunately, as we saw from the 1916 newspaper accounts, shark-attack stories have a tendency to be told, retold with great embellishment, and written with even greater sensationalism and with sketchy facts. The product is sometimes so mixed with fantasy and myth that it is difficult to salvage a meaningful word from the narrative or an ounce of confidence. In 1958, however, in an effort to chronicle all known and reliable shark-attack accounts, and to provide valuable source material for study, especially as to why such events occurred, the U.S. Navy, with the cooperation of the Smithsonian Institution, began compiling the International Shark Attack File (ISAF). The file is now administered by the American Elasmobranch Society with the Florida Museum of Natural History and is centered in Gainesville, Florida. The earliest recorded attack, as listed by the International Shark Attack File, took place in the year 1580, somewhere between Portugal and India, when a sailor tumbled overboard in rough seas. The unfortunate sailor managed to grab a lifeline tossed down by his shipmates and was hauled to "within half the distance of a musket shot" when a "large monster called tiburon [shark]" suddenly appeared from below and "tore the man to pieces before our very eyes. That surely was a grievous death."

Subsequent accounts among seafaring men are not rare, and in 1595 it was reported that:

this fish doth great mischiefe and devoureth many men that fish for pearles. . . . As our ship lay in the River of Cochin [India] . . . , a Saylor beeing made fast with a corde to the ship, hung downe with halfe his body into the water . . . and there came one of those Hayens [sharks] and bit one of his legs, to the middle of his thigh, cleane off at a bite, notwithstanding that the Master [Captain] stroke at him with an oare, and as the poor man was putting down his arms to feel his wound, the same fish at the second time for another bit did bite off his hand and arme above the elbow, and also a peece of his buttocke.

As indicated, in 1916, scientists and ichthyologists were unconvinced and practically unaware of reliable early attack records. Without thorough indexing and the ease of microfilm searches, the researchers of years past could have easily missed credible attack accounts from centuries earlier. The aforementioned sixteenth-century accounts would likely have been of some interest to the American Museum scientists in Manhattan, but if they had been acquainted with the 1809 edition of Diedrich Knickerbocker's *History of New York,* they may have even discovered the earliest recorded shark attack in the New World. I am indebted to Jim Foley, a friend and colleague, for unearthing Knickerbocker's account of an event from the 1640s, which is as follows:

It was a dark and stormy night when good Antony Van Corlear arrived at the creek [Harlem River] which separates the island of Manna-hata from the mainland. The wind was high, the elements in an uproar, and no Charon could be found to ferry the adventurous sounder of brass [Van Corlear was a trumpeter] across the water. For a short time, he vapored like an impatient ghost upon the brink and then, bethinking himself of the urgency of his errand, took a hearty embrace of his stone bottle, swore most valorously that he would swim across in spite of the devil, and daringly plunged into the stream. Scarce had he buffeted halfway over when he was observed to struggle violently battling with the spirit of the waters. Instinctively, he put his trumpet in his mouth and, giving a vehement blast, sank forever to the bottom.

Knickerbocker's account, however, does not end with the demise of Van Corlear. He goes on to state that an "old Dutch burgher," famed for his veracity and who had been a witness of the fact [Van Corlear's death], related the melancholy affair with the fearful addition that he saw the "dvyvel" [devil] in the shape of a huge fish "seize the sturdy Antony by the leg and drag him beneath the waves." The adjoining promontory, which projects into the Hudson, we are told, has been called "Spyt den Dvyvel" ever since and Antony still haunts the surrounding solitude as his trumpet has often been heard on stormy nights "mingling with the howling of the blast [lightning]." We are also informed that "nobody ever attempts to swim across the creek after dark . . . and a bridge has been built to guard against such melancholy accidents in the future."

Not only could Knickerbocker's narrative represent the first recorded shark attack on the continent, but because of the event's river location, it also provides an eerie prelude to the attacks at Matawan, New Jersey.

Gleaning hard facts from some shark attack descriptions (especially older ones) is sometimes frustrating, and any author who sensationalizes actual attacks certainly is no help to attack researchers and no help to sharks. Countless stories and negative references have certainly not been in the best interest of sharks. In actuality, except for the killer whale, the sperm whale, and some large sharks, modern man is the only predatory threat to sharks. Every year, fifteen hundred to two thousand sharks are found trapped in the nets off the South African coast, and thousands more fall prey to game and commercial fishing hooks around the world. To make matters worse, sharks have a low fecundity and do poorly in recovering from overfishing of their species. The white shark, for example, has almost no resilience to bounce back from fatalities that come from unnatural pressures. In South Africa, the white shark is now listed as an endangered species, and in California, all fishing for white sharks is prohibited with limitations also set for blue, leopard, thresher, and shortfin mako sharks. In 1996, the National Marine Fisheries Service proposed a ban on fishing for white sharks and basking sharks in the Atlantic Ocean and the Gulf of Mexico. The Australian government has granted official protection to Dangerous Reef, a tiny area off the southern coast known for white sharks.

The recent attempts at protecting the shark simply represent man's 180-degree turn toward understanding the shark and the predicament in which we have placed it and ourselves. In a way, we are saying of the shark, "We once thought you were malevolent and preferred to kill you off, and now we understand your natural magnificence, importance to the marine environment and your right to live."

When speaking and elaborating on shark attacks, accurate perspective is a key ingredient of a fair assessment. Richard Ellis and John McCosker, the authors of *Great White Shark,* put shark attack threat into concrete perspective when, after describing the slow moving or sedentary marine dangers such as stinging jellyfish, electric rays, and knife-edged corals, they wrote:

> Aloof from these passive threats of the reefs and the seafloors, who speak only when spoken to, there is another class of sea creatures, cruising the mid-depths, that are more hostile. These are the animals that look for trouble, that provoke confrontation, that attack. The most notorious of the attackers are the sharks, but ... [out of the 390 species of sharks] only about twenty are known to have attacked humans, and of these, only four—the great white shark, the bull shark, the tiger shark, and the oceanic whitetip—have done so on anything but the rarest of occasions.

During 1995, the worst shark attack year ever reported, there were seventy-two attacks. That number still comes to only 1.38 attacks per week around the entire world.

The whys and hows of shark attack research move to another level when particular shark species are analyzed for their contribution to the topic. Case studies, bite patterns, locations, and environmental conditions are some of the elements that are used in conjunction with an alleged species to make thorough sense of an attacker's profile and to suggest valuable reasons to explain the attack. Obviously, the species responsible for every or even most attacks cannot be definitively established and eyewitness descriptions of the offending shark are often clouded in exaggeration or lack important markers of a particular species' physical attributes. An experiment at the Mote Marine Laboraory in Sarasota, Florida, where subjects were asked to gauge a shark's size as it passed by in a tank, reported overestimations of large shark lengths in 66 percent of observations made by seventeen college students. Today, bite patterns, analysis of tooth fragments (which may have remained in the flesh or bones of the victim or an object), and the location of the attack are the three most helpful features that delineate the species recorded for an attack. Of these three factors, location and time of year are the pieces of information most available to determine a species.

With a growing data pool and with the speed and fluency of computer technology, organizations like the ISAF can provide attack species, locations, and trends at the touch of a keyboard. Such an accounting and sophisticated analysis can now define, dismiss, alter, or confirm theories and assertions that early attack researchers had proposed nearly half a century ago. The statistics relating to unprovoked attacks, through May 1998, reveal the top four contenders to be the great white (*Carcharodon carcharias*) with 231 attacks, the tiger shark (*Galeocerdo cuvieri*) with 67, the bull (*Carcharhinus leucas*) with 57, and the sand tiger (*Eugomphodus taurus*) with 31. Realistically, however, almost any shark in the size range of six feet (1.8 meters) or greater is a potential threat to humans. The top three sharks mentioned above are especially dangerous because of their cosmopolitan distribution and their large size, and because their diets regularly include large prey such as marine mammals, sea turtles, and other sharks. These species are likely responsible for most of the serious but less common "bump and bite" and "sneak" attacks.

The offending parties in the more common "hit and run" attacks are seldom observed in the act, but evidence from Florida, where most of these attacks occur, suggests that the blacktip shark (*Carcharhinus limbatus*), the spinner (*Carcharhinus brevipinna*), the blacknose (*Carcharhinus acronotus*), and the sand tiger or shovel-nose shark (*Eugomphodus taurus*) are the major culprits in that region.

In recent years, the tiger shark has become especially worrisome to the tourist industry in Hawaii. After a rash of attacks in the early 1990s, including a woman snorkling not far from her hotel in Maui, the initial response was to kill the offending culprit or culprits. That effort soon transformed into an ugly attempt to destroy as many tiger sharks as possible. Such an effort was met with great resistance from native Hawaiians who revere the tiger shark and hold it as integral component in their spiritual reincarnation beliefs. Over the past five years, the attacks have become much less frequent, but the tiger shark population does seem to have suffered from the initial overreaction The tiger shark is known for stripes across its back that fade in adulthood, but the bars often persist on the flanks or in the caudal region. The tiger is a powerful predator and its general coloration varies from bluish or greenish gray to black above and from light gray to dirty yellow or white below. Adults may grow up to eighteen feet. The tiger shark is especially abundant in the Caribbean Sea and the Pacific Ocean near Hawaii. Unfortunately, for the snorkler or surfer, the tiger shark does swim close to shore, sometimes venturing into bays and docksides. It is considered the "garbage disposal" of the shark kingdom and has consumed inanimate objects ranging from license plates to paint cans. Another attribute of the tiger shark, which is unusual for a shark of its large size, is its consumption of sea birds. During a two-week period in the northwestern Hawaiian Islands, these sharks ignore their usual prey items of seals, turtles, and lobsters to feast on albatross chicks that fall into the sea off cliff nests. The tiger shark is certainly deserving of its "man-eating" reputation, and it is no wonder that the scientists of 1916 placed it high on the list of possible culprits for the New Jersey attacks. It was also the tiger that was appropriately chosen by Peter Benchley as the "falsely accused killer" after the second attack in his novel *Jaws*. Who can forget the scene when Hooper (Richard Dreyfuss) pulls the Louisiana license plate from the tiger shark's insides?

The bull shark is another obvious contestant in the top three of human threats. It is now considered the most dangerous shark to humans in tropical and subtropical waters and grows to a maximum size of eleven feet. The bull's typical coloration is gray with an off-white underbelly. The bull has an upper jaw equipped with serrated teeth and a lower jaw of finely serrated, pointed teeth. The lower jaw acts to anchor the prey item while the upper jaw gouges out chunks of flesh. The bull, like most large carcharhinids (as well as the white shark), uses its powerful "torso" and head to violently jerk from side to side with a raking effect. The leverage that sharks produce with their tremendous bodies was the one factor (besides bite power) which the 1916 scientists seemed to underappreciate. The bull's huge jaws and impressive teeth come at the front end of a powerful body. As far back as 1948, Henry Bigelow and W. C. Schroeder, editors of *Fishes of the Western North Atlantic*, described the bull shark as a heavy, slow-swimming species, capable of swift movement when attacking. The bull shark has a varied diet. Large prey, including other sharks such as the hammerhead, dolphins, stingrays, and sea turtles are its main food.

According to Stewart Springer, author of the *Natural History of the Sandbar Shark*, the bull shark is notable among all sharks for its inclination to feed on young sharks of other species, particularly the young of its near relative, the sandbar shark. Springer states that the distribution of these two sharks is interconnected on this basis in that a female carcharhinid shark taken close to shore during the summer will almost always be a sandbar shark, while a male taken at the same time and place will almost certainly be a bull. Interestingly, female bull sharks seem to be inhibited from feeding while spawning, and since they are often near shore during this period, they have given many the very false impression that they are docile. Most attacks on bathers in South African waters are now attributed to bull sharks, and they have even been known to kill adult hippopotamuses in African rivers.

A quality of the bull that adds to its already competent potential to inflict injury on man is its unique ability to live in both fresh and salt water. It prefers shallow inshore areas, rivers, bays, and wharfside locations. The bull shark apparently has an elaborate system that allows it to physiologically acclimate to fresh water. Some bull sharks have been found hundreds of miles up the Mississippi River and twenty-three hundred miles up the Amazon. Since it frequents inshore waters and has a vast distribution, it commonly ends up in recreation centers. Since it does not have distinct external characteristics to the untrained individual and is not easily distinguishable from other sharks (especially the sandbar), it is often unrecognized as an attacking species. The bull shark is also known as the cub shark and regional variations of *Carcharhinus leucas* also include the Ganges River shark, the Zambezi shark, and the Lake Nicaragua shark (completely capable of surviving much of its life in freshwater). In *Sharks of the World*, Rodney Steel explains how the Lake Nicaragua shark is able to maintain a viable osmotic balance between its body and freshwater by adjusting its blood-urea content. It was once thought to be completely landlocked in Lake Nicaragua, but it is now known to negotiate rapids and return to the sea.

An early experiment by Dr. Perry Gilbert, of the Mote Marine Laboratory, demonstrated how innately cautious other intelligent species are in regard to the bull's prowess as a dangerous shark. Gilbert, in an attempt to train bottlenose dolphins to injure or kill sharks that may be threatening to man, routinely had a dolphin butt a lemon shark's belly with its powerful snout. When it came time to duplicate the act with a bull shark as the subject, the dolphin wanted no part of the experiment.

I mentioned earlier that technology, information gathering, and the advances of time will help sort out the mysteries regarding a species' predilection to attack. One scientific gray area has been identifying the species of the most prominent attacker in South African waters and the shark most responsible for the rare fatalities along the east coast of Florida, the Caribbean, and the Gulf of Mexico.

During the summer of 2000, an unusual sequence of attacks took place off the Gulf Shores area of Alabama. On June 9, Chuck Anderson, forty-four, a high school coach and vice principal, and Richard Watley, fifty-five, a barber shop owner, were swimming along the shore in training for a triathlon. Anderson, closer to shore, was viciously struck on the right arm and the right side of his trunk. He staggered out of the water and later required amputation of his hand and forearm, inches below the elbow.

Watley, about sixty feet from shore, did not know the fate of his partner but did notice that he was in some distress and merely thought he had encountered a jellyfish. The shark came right under Watley and then bit him in the thigh. Watley apparently prevented the shark from gouging out a chunk of flesh by repeatedly punching the shark. According to Watley, "I thought it might leave me alone, but it came at me again and again. I would punch him, he would retreat, and then I would swim as fast as I could for about five to ten seconds, but then I would have to turn around and face him again. He chased me all the way to shore." Two days after the attacks a twenty-two-foot boat transom was gnawed by an eight-foot shark in the same area of Alabama. ISAF director George Burgess attributed the first attack scenario to a full-grown, eleven-foot, five-hundred-pound bull shark, and another expert stated that the practice of long-line fishing in the vicinity of shore was a factor in drawing these dangerous large sharks toward recreational swimming locations. In August 2000, a man was killed in the Gulf of Mexico near Tampa by a shark thought to be a bull shark as well.

The designation of the bull shark as the culprit in the recent Alabama attack is a perfect example of a using the victim's vague description of a shark, the severity and character of the attack, and location of the attack to determine the species most probably involved. Even if a trained professional had observed the actual Alabama attack, it is classically difficult to discern the species of a carcharhinid shark during the heat of an attack. The tiger shark, for example, is a darkly toned, large, voracious species that is known to frequent the Gulf of Mexico during the summer, but for reasons that I am not fully clear on, it was not mentioned as a possible attacker. Additionally, the tiger has been known to attack multiple victims in the wild, while the bull has not.

In 1975, the *Book of Sharks*, written by America's foremost marine natural history author and painter Richard Ellis, was the first publication to describe the bull shark with the intensity and respect that it commands today. Ellis was also one of the first authors to comprehensively revisit an analysis of the 1916 attacks since the 1963 classic *Shadows in the Sea*. The senior author of *Shadows in the Sea*, Thomas Allen, provided Ellis with his voluminous 1916 source material for *Book of Sharks*, and Ellis concluded that a bull shark was the logical perpetrator in at least three of the infamous New Jersey attacks. In that discussion, Ellis, almost prophetically, states that the bull shark would someday be recognized as the most frequent shark threat toward man. Incidentally, Ellis's account of the 1916 attacks was also a major influence on this author to take up the torch and continue research.

The bull shark's propensity to attack people in freshwater rivers is indisputable, but its involvement in oceanic attacks is a bit less definitive. In the 1960s, Dr. David H. Davies implicated the bull shark in the majority of South African Natal coast attacks, while in 1975, A. J. Bass and J. D. D'Aubrey of the Oceanographic Institute of Durban wrote that while the bulk of attacks in Natal waters appear to be due to one or another carcharhinid shark, many of the attacks in the cape seas (Western South Africa) are probably the work of the great white. They also mention that small whites appear to be fairly common in southern and southwestern cape waters. In the late 1980s, Beulah Davis and her associates of the Natal Sharks Board (NSB) came to the conclusion that the great white and the tiger shark were the most aggressive and dangerous sharks off the Natal coast, and they considered those two sharks responsible for most of the fatal and near fatal attacks off the South African coast.

More than thirty years ago, not all icthyologists were convinced of the bull shark's man-eating prowess. In 1948, Henry Bigelow and W. C. Schroeder wrote, in *Fishes of the Western North Atlantic,* that the bull's reputation as a man-eater or man-killer may not be fully deserved "for otherwise shark fatalities probably would be far more frequent in Florida and the West Indies, where it is one of the more common of the larger sharks." In some respects, the Bigelow and Schroeder work is outdated, and since that time, Florida has become the unofficial shark-attack capital of the world. The premise of their comment, however, as it relates to attack *fatalities,* still holds true statistically. Of the 187 attacks in Florida over the past decade, only two fatalities were reported. George Burgess, therefore, qualifies Florida's new shark-attack title by more specifically designating Volusia County, Florida (the place where most of the attacks occur), as the "shark nip capital of the world" because of the minor nature of the majority of attack wounds. In India, however, the bull shark is vastly dreaded where it attacks pilgrims in the Holy River Ganges and feeds on corpses consigned to sacred waters.

Regardless of the debate over the bull shark's predisposition to inflict lethal wounds, it is certainly an exceedingly dangerous shark that has an infamous track record for causing severe injury to humans. Almost simultaneous with the publication of Richard Ellis's *Book of Sharks* in 1975, Grosset & Dunlap published *Sharks: Attacks on Man,* by George A. Llano. In the book, Llano also drew a parallel between some aspects of the 1916 attacks and that of the behavior of the Lake Nicaragua shark.

The third and most renowned of the top three attackers is the white shark. Commonly known as the great white shark (*Carcharodon car-*

charias is Latin for "the biter with the jagged teeth"), it is the largest of all carnivorous sharks. Adult size ranges from sixteen to twenty-one feet (but unconfirmed reports have stated lengths up to thirty feet). Adult great whites generally weigh between one and three tons. Such dimensions would place the large adult specimen equal in size to a full-grown rhinoceros. Its upper and lower jaws are uniquely equipped with heavily serrated, triangular, razor-sharp teeth. It is found in every ocean on the planet, but it is not abundant in any location. The white favors cool, temperate waters such as the offshore waters of southern Africa, southern and western Australia, the Pacific Coast off northern California, and the offshore waters of northern New Jersey and New York. It is known as the white pointer in Australia and the blue pointer in South Africa. Its color variations come in slaty brown, dull blue, gray, or even charcoal black above, and dirty white below.

Many marine biologists consider the white shark the most danger-ous fish in the ocean, and more attacks are attributed to it than to any other species. Still the most feared of all sharks, it has been known to attack boats, sometimes persisting until the boat sinks. It has even been known to leap into boats. The white shark accounts for the vast majority of attacks on humans off central and northern California, southern Australia, and the Cape seas of South Africa.

Acoustic telemetry experiments have revealed that the great white shark can travel more than a hundred miles in a three-and-a-half-day time span. Such a revelation about the white shark's long-range cruis-ing ability may be linked to its internal temperature. In the early 1980s, the late Frank Carey, an ichthyologist at Woods Hole Oceanographic Institute, put a temperature probe in the muscle of a great white and found that the shark's body temperature was higher than the sur-rounding water. A few years later, John McCosker, an ichthyologist at the California Academy of Sciences, induced a great white to swallow a temperature probe hidden in a slab of seal blubber and confirmed Carey's discovery. Like only the porbeagle shark, the mako, and the tuna, the great white is warm-blooded. Its ability to regulate body temperature accounts for a differential of at least 15 degrees Celsius above the surrounding water. A warm core temperature allows the great white to cruise on extended hunts, make numerous dashes at prey, and participate in furious combat with powerful marine mammals.

The fact that great whites are warm-blooded may also account for their dietary choices and attack behavior. White sharks normally feed on a variety of bony fishes such as tuna, but they also feed on other sharks, dolphins, and an assortment of pinnipeds such as harbor seals, sea lions, and elephants seals. They have also been known to feed on garbage, dead animals, and even the sea robin (*Prionotus* sp., dreaded by fishermen because it is spiny and virtually inedible) off the New Jersey coast.

The white shark diet and its formidable reputation may lead one to believe that it is an indiscriminate eating machine. However, larger white sharks seem to favor fatty prey items because of the need to burn prodigious amounts of calories. We have long known that the huge predatory polar bear prefers adult seals to newborns because of the high fat content of the mature individuals. In reference to oceangoing creatures, researchers have postulated that the energy needs of warm-bloodedness could place great demands on an animal that lives in cold waters.

The fact that one gram of fat contains nine calories, compared to four calories per one gram of protein, explains why Dr. Peter Klimley, of the Bodega Marine Laboratory, considers the fatty elephant seal the "PowerBar of the sea" for the white shark. In reality, recent research has determined that the white shark is a relatively "picky" eater with a sophisticated feeding system. In one the most productive fisheries on the West Coast, near the Farallon Islands, some thirty miles west of San Francisco, video recordings of 129 great white attacks on seals and sea lions have been examined. An analysis of this footage (some of which shows seals being violently torn apart and others being decapitated) suggested that a shark near the surface would bite its victim, drag it down bleeding, carry it underwater, and perhaps take another bite, then let the carcass float upward. If the victim was still alive, the process might be repeated. The scientists also observed that the great white would take a tentative bite of an unfamiliar object, such as a buoy or a surfboard, then spit it out. The whites also gingerly bit a fake plywood seal and tended to initially "mouth" prey candidates delicately rather than just munch. One scientist even suggested that the shark's have a "soft mouth," like bird dogs, and that they get a tremendous amount of information from the "gentle mouthing process." The shark's aim, the scientists deduced, was to "check out" the prey item, bite forcefully, then have the prey bleed to death as soon as possible, minimizing a potentially violent and dangerous struggle.

Dr. Klimley, one of the three scientists who worked on the feeding study, noted that the feeding pattern exhibited with the seal/sea lion attacks was similar to the pattern displayed in attacks on people. With the human attacks, however, the shark would often let the victim go (get away) after dragging him down. This behavior is also reflected in the fact that only 10 percent of the California white attacks have resulted in death to the victim. Some researchers originally felt that white sharks might find human flesh unpalatable. Dr. Klimley believes that white sharks might "spit" out humans, birds, and sea otters because their bodies lack the energy-rich layers of fat that their preferred foods (like seals and whales) possess. Otters and penguins have been injured and even killed by great whites, but for some strange reason they were not consumed. It would be silly to conclude that the whites were waiting for the animals to bleed to death before consumption, and then forgot about going in for the meal. In other words, the theory goes, if the tensile resistance associated with blubber is revealed, the shark goes for a full-strength bite and eventually swallows the prey material. If not, it will back off to save its energy for a more nutritious meal. If a shark were to occupy its digestive tract with a large nonnutritious/poorly energy-efficient meal, that object might take up the room of a potential prey item with more abundant calories. Klimley also believes that this behavior is the reason why most humans who are bitten are seldom killed (finished off, consumed). The ratio of muscle to fat is simply too high for a white shark to find us acceptable for consumption. Sea otters, which have dense pelts rather than fatty blubber, are, in fact, sometimes found floating dead with white shark teeth fragments embedded in their flesh.

Not all experts agree, however, with Klimley's hypothesis. George Burgess, for example, notes that, in a third of attacks on humans, the white shark returns for a second bite. Burgess stresses that more victims would become shark food if it were not for speedy rescue intervention. As far back as 1986, in *In Search of the "Jersey Man-Eater,"* I presented my views on the general "quick release phenomenon" exhibited in some attacks. I mentioned that the shark's drive to feed may actually be "turned off" by the unexpected bony frame of the victim. I also noted that factors such as "fear," agitation, or territoriality might be behind attack actions, as opposed to intended consumption. Most interestingly, I stressed the observation that Dr. Perry Gilbert had made in referring to the shark as an "opportunistic feeder." That is, they tend to prey on individuals who are weak(ened), injured, or off guard. The white shark, for example, usually attacks its prey by surprise, moving ten to fifteen miles per hour, then suddenly rushing its prey from behind and below. It would not be surprising to find that a shark might view a swimmer's erratic/unusual movements (e.g., splashing, jumping, choppy stroking, static wading, deliberate motion) as a distress pattern or a sign of being unprepared. Such a conception would signal a forceful attack, a test bite, or a "look-see." In *"Jersey Man-Eater,"* I also suggested that, in an attack by a shark that is less than ten feet in length, a victim's surprisingly vigorous response in combination with his unusual taste, bony texture, and long-limbed (nonfatty and nonflippered) structure may influence a shark's quick release.

One factor that may make white sharks especially predisposed to a cautious preliminary attack pattern (this is relative, of course) is the nature and importance of their vision. The black, circular, "pupil-less" eyes of a white shark certainly add a sinister aura to the beast's already terrifying appearance, but the great white's eyes are also distinguished by other anatomic and functional features. The *Carcharodon* eye possesses both rod and cone photoreceptors, but the cones are arranged in decreasing numbers at the periphery of the retina. Such an arrangement suggests that its vision is poor in dim light, and its nocturnal vision is poorer still. The relatively small size of the white shark's eye, as well as other internal anatomic features, supports the white's tendency toward feeding at the illuminated surface. An internal optical "clear layer," called the tapetum lucidum, which utilizes a series of mirrorlike surfaces to augment ambient light, is not as well developed in the white shark as in other species. This further requires the white's dependence on daylight.

The white shark is the only shark that lifts its head out of the water to look at objects on the surface. The white shark's tendency to feed at the surface is apparently so great that abalone divers have actually saved themselves from aggressive whites by remaining low to the bottom for up to four hours using hookah rigs.

Marine biologist John McCosker has even found that whites avoid diver dummies that stand at the bottom as opposed to the ones that float at the surface. Besides providing needed light or clear water, the surface may also provide added protection against potential eye harm. The bottom of the sea is congested with pointed objects such as branching coral, rock formations, ship wreckage, pilings, refuse, etc., and such obstructions or protruding objects could cause great damage to the vulnerable eye of the violently feeding white or at least hinder its vision (e.g., in kelp beds). Unlike some sharks and assorted other animals, the great white eye lacks a clear, protective eyelidlike nictitating membrane beneath and separate from the outer eyelid. Instead, a white shark's eyes roll back in its head for protection exposing a white opaque covering when any contact with prey is made or when any collision arises. Some authors have suggested that this momentary blindness allows some agile prey to escape. Richard Ellis stresses that the white shark's only vulnerable area is its eyes (although the gills may be vulnerable as well) and has hypothesized that the seal's best defensive gambit, if attacked, is to claw at the shark's eyes.

Rodney Fox, the famed Australian great white attack victim, actually triggered the ultimate release of his attacker by gouging at its eyes. Those susceptible eyes, therefore, may be an added reason why birds, sea otters, and human beings are not usually fully consumed by the white. It is puzzling, however, that the white shark does not hesitate to engulf a barb-tailed stingray and subject its digestive system to the indigestible knifelike tail barb. The difference is that the tail of a ray lacks the finely tuned dexterity of mammalian or avian appendages. In other words, it apparently considers the pointed talons, quill feathers, and beaks of birds in the same unpalatable category as the claws of sea otters and long bony limbs of humans.

While some sharks are incredibly fast swimmers, great whites normally require the element of surprise to seize prey that may be quicker or more agile than themselves. When not actually in the act of feeding, the white will swim below the surface to stalk prey above. The silhouette of the prey contrasts against the bright sky. The white shark below, on the other hand, camouflaged by its dark back, is able to rush up at its prey without any early detection. Sometimes it even exits the water to seize its target. Whites have been known to snatch resting seals and sea lions off rocks, several feet above the surface of the water. Still, each of these predatory tactics requires a well-honed use of vision. Therefore, when it comes to attacking a human being, a shark may be able to instinctually weigh the reward against the danger and decide if the attack is "cost effective."

The white shark's particular dependence on vision for daytime predation, as well as the fact that its eyes represent its only "Achilles' heel," may explain its sometimes ginger biting habits and its frequent aversion to consuming a human being. We possess all the features of an unpalatable, poorly efficient, and risky dietary choice. We are large, bony, vigorous when aroused, and intelligent. To a white shark, the permanent loss of its full visual acuity may be akin to a blue crab loosing both its claws or a swollen-faced boxer losing the vision in one eye. The fleeting moments that make up the time span of most attacks are apparently more than enough time for the white shark to determine that man is just not worth the effort. Other researchers refute the possibilities that sharks (at least the white shark) are inefficient feeders. They maintain that shark may not, in fact, be desperate at all in their quest for the right food and may actually be displaying agressive behavior rather than feeding.

Regardless of the debate over why the white shark does not fully devour every one of its victims, and despite the increasing attack notoriety relating to bull sharks, the great white is still the unchallenged master of seaside terror. As Richard Ellis wrote in *Monsters of the Sea,* "The great white is firmly ensconced in the pantheon of sea monsters." Other authors, scientists, and residents of assorted regions have referred to the white shark as "white death," "death shark," "maneater," "the ultimate predator," "the most formidable fish-like vertebrate," "a killing machine," "the last free predator of man," "the one shark that fears nothing," "the most feared silhouette in the underwater world," "the most terrifying creature of the ocean," and "the most frightening animal on earth."

One Spanish designation for the white shark, "*devorador de hombres,*" even sounds reminiscent of the Spanish term used to describe the infamous alien hunter from the Arnold Schwarzenegger film *Predator* (when translated, it was the "*Predator,* which made trophies of men.") Perhaps the most poetic encapsulation of "white terror" comes from none other than the story of the white whale, *Moby Dick.* As Richard Ellis and John McCosker, the authors of *Great White Shark,* appropriately cited, Herman Melville captured the essence of the white shark's preeminent predatory aura when, in 1851, he wrote:

This elusive quality it is, which causes the thought of white-ness, when divorced from more kindly associations, and cou-pled with an object terrible in itself, to heighten the terror to the furthest bounds. Witness the white bear of the poles, and the white shark of the tropics; what but their smooth, flaky whiteness makes them the transcendent horrors they are? That ghastly whiteness it is which imparts such an abhorrent mildness, even more loathsome than terrific, to the dumb gloating of their aspect. So that not the fierce-fanged tiger in his heraldic coat can so stagger courage as the white-shrouded bear or shark.

Although the International Shark Attack File reports the earliest re-corded shark attack to have occurred in the sixteenth century, an ear-lier known attack by a great white may have been overlooked. This at-tack concerns a victim named Jonah, the biblical prophet who lived from 786 to 746 B.C. That Old Testament text, written between the fifth and third centuries B.C., provides a didactic narrative depicting the interaction between innocence and evil. The story's first and sec-ond chapter describe an initially disobedient Jonah who departs from the port at Joppa to undertake a long sea journey. Not long after the journey begins "the Lord, however, hurled a violent wind upon the sea." Jonah, believing his presence was the reason for the Lord's wrath, volunteered to be thrown overboard. The panicked crew, now rowing briskly for shore, gladly obliged Jonah's offer. When Jonah struck the water, the rough sea suddenly abated. From childhood, many of us know the story to follow with Jonah's encounter with a hungry whale. In actuality, after Jonah's plunge into the violent waves, the text reports that "the Lord appointed a large [or great] fish to swallow up Jonah; and Jonah remained in the belly of the fish three days and three nights." The question arises: Was the Old Testament author here merely using the term "great fish" to describe a whale, or has subtle misinterpretation been perpetuated through the ages? And if the traditional whale were really a shark, would the great white species be a logical choice because of its large size and its reputation as a man-eater/man-biter?

The earliest written reference I could find to Jonah's close encounter with the sea creature comes from the Bible itself. In chapter 12, verse 4, of the Gospel of Matthew, written between 40 and 70 A.D., the writer here refers to the culprit as "the whale." According to *The Shark Almanac,* a legend from Papua New Guinea tells of the miraculous deliverance of a local tribesman named Mutuk, a man who was swallowed by a "shark." Mutuk's story and that of Jonah are basically the same. In the sixteenth century, French naturalist Guillaume Rondelet, who, according to *Great White Shark,* may have produced the earliest illustration of the white shark, is also the first scientist to suggest the great white as the likely swallower in the story of Jonah. Rondelet also reported finding a fully clad intact sailor in the stomach of a great white. In the *Systema Naturae of 1758,* written by the Swedish naturalist and physician Carolus Linnaeus, the animal classification text mentions that the white shark was an enormous fish and probably the variety that swallowed Jonah. Bishop Erik Pontoppidan of Norway, a prolific writer of seagoing fiction and nonfiction, in 1765 wrote a detailed paper that proved, to his satisfaction, that Jonah had been gulped down by a basking shark. In 1916, some scientists revisited the question of Jonah's attacker as it related to the New Jersey attacks. (It is no surprise that the astonishing impact of the epichol attacks would compel some to think in biblical proportions.) Paul Budker, in the *Life of Sharks,* had reviewed the early writings of Rondelet and concluded that Rondelet felt it impossible for a man to pass down the narrow throat of a whale and that the *Carcharodon* was not a bad choice because of its ability to swallow large prey whole "and bring it up later." Additionally, sharks have been known to regurgitate the contents of their stomachs at will. The only other marine animals in the running for such a position would be the sperm whale, the killer whale, the whale shark, or the basking shark, although a grouper has also been implicated. Budker, in his certainty about the assignment, even felt that *Carcharodon carcharias* should be specifically substituted for "whale."

A closer look at the "style" of the Jonah "attack" story, through the eyes of twenty-first-century shark knowledge, doesn't put us much closer to proving or disproving the great white assertion. Generally speaking, the Jonah incident might be the earliest known example of a "bite and spit" attack (with a delayed spit). What can be scrutinized, however, is the location of the action. According to the story, Jonah was not far from the shore of Joppa when the incident occurred. Joppa, in ancient times, was the original port of Jerusalem on the Mediterranean Sea coast of Israel. The fact that Joppa was also a whaling port in ancient times may have influenced the scribe of the Gospel of Matthew to choose the whale for his writing. On the other hand, the very fact that whale carcasses may have been in the general vicinity of Joppa additionally supports the idea that great whites were in the area. Great whites have been known to ravage the floating remains of whales off Montauk, New York, and, in view of the whale's blubber (fat) content and its passivity (dead), the whale carrion likely represents the "filet mignon" of white shark dietary opportunities. But could a great white attack occur in the region described by the biblical writers? The answer seems to be yes. Since 1876, there have been twenty-two confirmed unprovoked white shark attacks with thirteen fatalities in Mediterranean waters. In 1891 (from *Mediterranean Naturalist*), at least one Mediterranean scientist, in his assessment of natural influxes of fauna that arrived with increased temperature and humidity, stated that "it is not unusual to find the Great White Shark of the Indian Seas disporting itself in the waters of the eastern basin (through the Suez Canal) of the Mediterranean." In the end, it appears that the Gospel of Matthew may have been the first to formally interpret "great fish" as "whale," and it was that early interpretation or nomenclature, along with our own powerful inclination to assume automatically that a man would not be able to *gently* inhabit the alimentary apparatus of any sea creature other than a whale, that promoted the story as we traditionally know it. The significance of "righting" the identification of the Jonah monster would obviously be of interest to the shark-attack researcher and the historian, but for all intents and purposes, the designation of the creature as a "great fish" or a "whale" has no philosophical or theological significance. The Jonah beast is not the principal item in the story, only an obedient agent of God's purpose.

The concept of multiple attacks and rogue-type attacks should certainly be examined in any preliminary discussion of the New Jersey attacks of 1916. I made brief mention in the Introduction of frenzy-type feeding among sharks. That behavior constitutes a situation whereby competitive action toward prey items becomes standard. What initially begins as a deliberate and organized approach, in which individual sharks single out one prey item, turns into an all-out indiscriminate barrage of shark biting at any object or item in sight. Even other sharks become victims. This behavior is not limited to sharks, and it can even be seen in human mob/riot situations where sensory overload disrupts normal inhibitions. Outside of feeding frenzies, however, a shark does not usually direct its attention or aggression toward multiple objects or victims. Such a phenomenon may benefit the people who find themselves in the vicinity of an attack, and it may also makes the efforts of an attack rescuer much safer. According to H. David Baldridge's statistic-packed book *Shark Attack,* attacks on rescuers or multiple victims are extremely rare. In only 1 to 3 percent of attacks is a rescuer injured by the attacking shark. Among the most dangerous sharks, the only identifiable species to exhibit this type of multiple-target behavior are the tiger shark and the white shark. Up until 1986, I had only found one incident when a bull expressed such actions. In a Florida aquarium, a bull shark was provoked by being netted and turned and bit two handlers. The Gulf of Mexico attacks (presumably perpetrated by an eleven-foot bull shark) in June 2000, however, if confirmed to be a bull attack, would reveal that a bull is quite capable of pursuing two victims during the same attack incident.

Apparently some shark species are unable to redirect their aggression away from the target they originally singled out. One should recall that a shark is an "opportunist" that prefers to screen its prey to identify the most vulnerable single subject. This is one of the reasons why the experts suggest never swimming alone.

Case number 523 in the International Shark Attack File reveals the great white's bizarre potential for multiple attacks. Outside a reef in Southern Australia, searchers were attempting to hook into the remains of two men who had drowned the day before. The searchers spotted at the surface a white shark that quickly dove and then resurfaced with the body of one of the drowning victims. The shark then snapped at the searchers' grappling irons and persistently attempted to regain possession of the body even after it had been secured by the workers.

In case number 1247, a diver was surfacing with a speared fish in Australian waters when he was attacked by a very large white shark. The wounded diver made his way back to the dive boat and dropped the speared fish. At this point, the shark ignored the fish and turned its attention toward five other divers on the bottom. With two of the divers taking sanctuary in an underwater cave, the shark alternated its attention between the cave and the other three divers for several minutes.

To make blanket statements about a shark's potential behavior or predisposition based merely on its species may not be altogether accurate. It is my opinion that a given shark species in one area of the world may behave significantly differently than the same species or subspecies in other regions. This type of variation is seen in other predators as well. Most killer whales, for example, will limit their diet to fish and other small sea creatures. The "transient" killer whales off Mexico, however, tend to drive their enormous bodies on to the beach to snatch seals. Young orcas in this region even practice this feeding method by grounding themselves just beyond the surf. As mentioned earlier, some great white sharks forget that the textbook says that most white sharks are supposed to halt an attack after the first "distasteful" bite of a human being. After hearing of attacks off the coast of Chile in 1981–82 where two people were literally consumed by white sharks, John McCosker surmised that there might be varying races (subspecies) of great white sharks that could be less inhibited by the unusual human prey item. An attack in 1971 may also show that white sharks do not always have an automatic shutdown in feeding as a response to lean human flesh. According to Tim Wallett's graphic and informative publication *Shark Attack and Treatment of Victims in South African Waters,* a man by the name of Theo Klein was repeatedly mauled by a nine-foot white for up to twenty minutes. Similarly, the bull shark of India, South Africa, and Central America appears to attack with a greater degree of viciousness than the bulls of the western North Atlantic, the Gulf of Mexico, and the Caribbean. Among individual species of animals there has always existed subtle anatomic variations as well as variations based on the availability and varieties of prey fauna. If an individual species in one area of the world has developed a familiarity with hunting and dealing with a particular habitat and environment, then why would it be surprising to see subtle differences in the disposition of geographically separated species? It is usually inappropriate to compare a complex species like man to other members of the animal kingdom, but for the sake of clear comparison, consider the general demeanor of people who hale from different regions of the globe or even different states and cities.

As I alluded to earlier, I believe that research and writing about bull sharks from the 1960s and 1970s, especially from Richard Ellis and David Davies, has opened the eyes of many shark enthusiasts to the potential viciousness of bull sharks. Today, one cannot open a well-researched shark-related book without recognizing the agreement among shark authorities on the infamy of bull sharks. Since bull sharks are common in tropical waters and inhabit locations that are well traveled by recreation seekers, their reputation as a frequent

threat to man is understandable. In a head-to-head comparison between the bull shark's tendency to attack and inflict serious injury on
man and that of the great white shark, I believe that there is still no
contest. Just recently, I spotted a television program on bull sharks in
which two divers set out to get a never-before-seen close-up look at the
bull sharks of the Florida Keys. Much of the live shots and dialogue
seemed to be presented for shock value, but the video-equipped
snorkelers did obtain spectacular footage of up to thirty large bull
sharks cavorting with big tarpon just under the channel bridges. One
of the divers even grabbed a dorsal fin of a bull just prior to some feeding excitement brought on by a dead stingray. The program also
showed a moderate-size bull shark, frustrated in shallow water, ram-
ming its head into the transom of a boat. In the end, the divers referred to bull sharks as the most dangerous sharks on the planet. My
question is, Would these men even have attempted to obtain the
footage in the manner described if they were among white sharks?
And when I see research crews swimming among sharks like the blue
shark, testing protective steel mesh dive suits, I often wonder whether
a living diver would ever have the nerve to test such items, out in the
vulnerable open, against a great white. Earlier I mentioned the experiment that Perry Gilbert devised to train a dolphin to inflict injury to a
shark on command, and that Dr. Gilbert found the bull shark to be
quite daunting to the dolphin. I would venture to say that the same
dolphin in training, if presented with a white shark, would likely have
jumped out of the tank completely. This is no joke. On more than one
occasion, wild seals have been known to leap into motorboats when
being chased by a white shark. The desperate seal in each case obviously chose to take its chances with the human threat rather than with
the great white. One should also consider that the white shark is not
abundant in any area of the world. If the white shark were as common
as bull shark in a given region, it's quite probable that the whites
would pose more of a threat to man in that locale.

A discussion of shark attack in the context of an analysis of the 1916 tragedy would not be complete without a look at rogue sharks. Over the years, with man's increasing interaction, exposure, and encroachment upon natural habitats, certain powerful animal species, such as the elephant and tiger, have had individual specimens among them labeled as "rogues." This term is applied when one of the species behaves in an unusually vicious, persistent, or destructive manner, and probably relates to "rogue's" usage when describing someone who portrays a scoundrel, a vagrant, or a mischievous person.

Following his analysis of several shark attacks, including those off Australia in the late 1920s, surgeon and shark-attack maven Sir Victor Coppleson theorized that a shark (the rogue type), after consuming the flesh of a certain species of prey (i.e., man), will intermittently continue to seek out prey of that same species. Many sharks, including bulls, whites, and tiger sharks, have been implicated in rogue-type attacks with an average of two or three victims over a period of months and within a distance of thirty miles. These rogue series are said to end abruptly when, it is assumed, a rogue shark moves on elsewhere or with the capture of the shark (sometimes containing human remains) within the attack area. The average length of the attacking sharks (when described) is 6.6 feet. Not surprisingly, many, if not most, experts have criticized rogue-shark designations or viewed applications of the rogue-shark theory skeptically as sensational attempts to link unrelated attack incidents.

Notable sequential attacks that took place in Australia in 1929 and South Africa in 1957–58 and 1961 serve as a fair comparison to the description of the notorious American attacks that took place earlier in the century. Rodney Steel, in *Sharks of the World,* noted that Australia has always been the focus of shark attacks, with over four hundred deaths (the world's most), but South African beaches probably have the most notorious record of maulings with some sixty shark incidents along the Cape of Good Hope and Natal from 1925 to 1975.

An increasing urban population in the 1920s and 1930s led to widespread use of Australian beaches and resulted in regular reports of shark attacks. In the teens and early twenties, the people in areas like Sydney were becoming all too familiar with shark-related injury. At the fashionable Bondi Beach in Australia, on January 12, 1929, a boy was seized in only four feet of water and died after admission to the hospital. His wounds included a massive laceration from the top of his right hip to the middle of his thigh, the crest of the ilium (the back of the hipbone) having been torn and fragmented. On February 8, 1929, a thirty-nine-year-old man died before reaching Bondi's St. Vincent Hospital. He succumbed to a huge bite on his right thigh (fifteen by seven inches), which severed the femoral artery. Within the month, on February 18, at Maroubra Bay, Sydney, a twenty-year-old youth was savaged at 3:30 P.M. Extensive lacerations to his thighs, abrasions of the right leg and foot, and severe damage to his left hand were vigorously cared for. The man was given gas-gangrene antiserum, and his wounds were cleansed, irrigated with eusol (chlorinated antiseptic), and dressed. The young man died of sepsis within a week. The human toll in this potential rogue sequence was three dead in just under one month. No single shark was ever implicated in the attacks. Australia was not immune to another attack in that year, and although the time lapse does not seem to connect it as a rogue-attached event, its peculiarity makes it worth mentioning. On December 26, 1929, in the Parramatta River in the Sydney area of Australia, a sixteen-year-old boy was repeatedly attacked by a persistent shark that is said to have continued to maul its victim even as rescuers lifted the boy into a boat. As the boy was hoisted, the men recognized that his left arm was severed above the elbow and his chest suffered multiple lacerations. The youth died before he could be transported to a hospital.

The Zambezi shark (bull shark) of South Africa was implicated in three attacks that occurred within three months in the vicinity of the Limpopo River in January 1961. One hundred sixty-five yards up the river a child was mauled. The following May, a swimmer was killed in the same area, and, a few days later, a man sustained serious injuries when a shark struck at him in shallow water. The shark was never apprehended. By 1937, large-meshed gillnets were anchored just seaward of the breakers at many Sydney beaches. The nets not only trapped large sharks but also reduced the incidence of attack.

The beautiful beaches, warm water, and subtropical climate of South Africa's Natal coast (currently called KwaZulu-Natal) have long attracted tourists and residents to the sea. Like Australia's early experiences with attacks, the Durban City Council was concerned about the threat of shark attack as early as 1904. At that time, the city decided to erect a large semicircular enclosure, approximately one hundred meters in diameter, to protect swimmers. The rough surf finally had its way with the barrier, and by 1928 it was demolished. Durban was virtually attack-free until the late 1940s, when seven fatal attacks occurred over a six-year period. By 1952, Durban adopted the Australian gillnet method, and in the first year of operation, 552 sharks were snared and no serious shark-inflicted injuries occurred. Unfortunately, this atmosphere of safety did not prevail at other holiday resorts on the Natal coast, particularly those south of Durban.

In a bloody attack spree that falls second in infamy only to the 1916 New Jersey attacks, the Margate Beach area of South Africa, fronting the Indian Ocean, saw four shark attacks in one month. In December 1957, a Wednesday-Friday-Monday-Monday sequence of attacks had a devastating effect on the coastal tourist industry resulting from a mass exodus of panic-stricken holiday makers. That month of December would later be known as the infamous Black December, and in response to the public outcry, and fearing financial disaster, several coastal towns attempted to erect physical barriers in the surf zone. The unsightly structures, built from poles, wire, and netting, could not stand up to the heavy wave action and were soon abandoned. Depth charging by a South African Navy frigate is known to have killed eight sharks, but probably attracted more sharks to the area to feed on dead fish. Locals reported seeing huge live sharks being grounded in the surf as waters receded between waves.

The South African attacks all began when sixteen-year-old Bob Wherley lost a leg to a shark on December 18 at Karridene. Thirty miles south, at Uvongo Beach, Allen Green was attacked and killed on December 20. In George A. Llano's book, *Sharks: Attacks on Man*, C. J. Van de Merwe described what happened to James Vernon Berry on December 23, at Margate, a few miles from Uvongo:

> About 150 people were swimming. We were on the outer fringe in waist-deep water at about 4:30 P.M. We were 30 yards from shore when I saw a black fin the waves less than 15 feet away. I thought it was a snokeler's fin and didn't think much of it. Suddenly it appeared again and I was horrified to see that it was a 10 foot shark. I shouted to warn the other bathers and just as I turned to leave the surf I saw Mr. Berry get attacked as he was floating. It was too horrible to watch as the shark tore at him with his huge jaws while he tried to fight it off with his bare hands . . . his arm torn to ribbons to the bone. Again the shark attacked with vicious bites, tearing the flesh from the man's side and thigh and shaking him like a cat does a mouse.

Berry died in the ambulance on the way to the hospital. Perhaps the most horrendous of all the attacks took place on December 30, again at Margate, with the ocean filled with bathers. While a shark tower was being constructed, and a spotter plane was making rounds, a young woman named Julia Painting was viciously struck in the surf. Paul Brokensha went to her aid and punched at the shark but was initially thrown down by its tail. Painting was bitten in the presence of hundreds of beachgoers. Her left arm was gone, her breast and buttock were savaged, and her right hand was severely lacerated in attempts to defend herself. On the beach she was reported to say, "Let me die, let me die, I'm finished." Julia did survive and left the hospital in less than a month. Aubrey Cowen, the Margate lifeguard who assisted the victim out of the surf, stated: "Margate was a small town that depended on the tourist industry. When Julia Painting was at-tacked, that was like pulling the plug out of the bottle. Margate turned into a ghost town. There's nothing more horrifying than see-ing a victim of a shark attack."

There were suggestions that lard be placed around carbide balls. When swallowed by the shark it would form a gas and float the sharks to the surface. The SPCA, however, stopped the effort by calling it un-necessarily cruel. On January 9, at Scottburgh, between Karridene and Margate, Derick Prinsloo, forty-two, was swimming at an area where swimming was not banned. He was attacked in no more than two feet of water, and the shark continued the attack until its belly was on wet sand and Prinsloo was five yards from shore.

Four bathers were attacked in twenty-two days with two fatalities in a freakish killing spree. It was widely remarked that each of the victims wore bright orange, yellow, or red coloring in their bathing attire, and many wondered whether that fact contributed to their demise. With two more attacks occurring during Easter 1958, Margate had no alter-native but to install gillnets, as had been the practice at northern beaches and in Australia. Durban had actually restricted swimming to enclosed areas since 1916 (note the date). South African lifeguard George Plowman lost his right lower leg (below the knee) to a shark while bodysurfing in 1951. He recalls that the shark had his leg for only seconds, but when he looked down, his lower leg was gone. Plow-man also recalls that when the first five nets were placed around Mar-gate after the 1957–58 tragedies, 310 sharks were caught, including a sixteen-foot great white.

By 1964, the Natal Provincial Administration created a statutory board known as the Natal Anti-Shark Measures Board, now called the Natal Sharks Board (NSB). The NSB was "charged with the duty of ap-proving, controlling and initiating measures for safeguarding bathers against shark attacks." Originally, the Sharks Board was formed to su-pervise installation of gillnets at the larger holiday resorts to the north and south of Durban. The nets were maintained by commercial fisherman or municipal employees until 1974 (presumably after an attack at Amanzimtoti Beach), when the NSB began taking over the servicing and maintenance of the net operation.

The NSB nets serve as fishing devices and now span forty miles of coastline. Since sharks need to propel themselves forward to bath their gills with oxygen-rich water, being caught up in the net results in an 85 to 90 percent mortality rate for the sharks. These gillnets have a top to bottom width of twenty-three feet and are placed in water depths of thirty-two to forty-five feet. Such dimensions reveal that some sharks are able to swim under, above, and around the nets, yet by diminishing the relative number of inshore sharks, the number of attacks has dropped to a spectacularly low level. Amanzimtoti, once referred to as the most dangerous shark-attack vicinity in the world, credits the gillnets with keeping it attack free since 1974. Dr. David H. Davies, the former director of the Oceanographic Research Institute in Durban, was always perplexed as to the reason behind the gillnets' effectiveness as a shark deterrent even though it did not make a complete barrier. Interestingly, New Jersey had a correlate of the gillnets already in place in 1916. These were known as the pound nets. They were placed about a half mile from shore and had smaller openings than the gillnets, because the objective was to keep small "pound fish" from exiting. James Meehan, the Fish Commissioner in Philadelphia in 1916, was even hoping that the perpetrator(s) of the 1916 attacks would entangle itself in the nets. The holes in the pound nets were likely too small for that or for entangling any significant number of the larger inshore sharks of that year.

Obviously, lowering the inshore shark population is key to the success of the modern gillnets; in addition, shark fishermen have long held that the smell of decomposing fish, especially that of sharks, drives off sharks. Apparently the acetic acid content of the decaying flesh is the key, but with scores of sharks in the nets each morning, it seems unlikely the nets and rotting flesh repel the sharks. Copper acetate (derivative of acetic acid) was actually used as "Shark Chaser," the U.S. Navy's shark repellant in the 1960s. As the U.S. Navy dive manual reported, however, "shark repellants are useless" when sharks "are hunting in packs and food or blood is present." Perhaps the nets' display of decaying and disabled sharks is just enough chemical and visual deterrent to momentarily "turn off" a shark's appetite, or at least prevent it from being interested in a prey item as unappealing as a human being. The nets are checked every morning, and most live sharks (the few that survive) are set free and tagged, including the 150 specimens of the whites, bulls, and tiger sharks caught annually. The dead sharks are carefully weighed, identified, and dissected. All South African attacks are reported to the NSB, and the information is then conveyed to the ISAF.

The circumstances surrounding the 1916 attacks clearly reflect the shortcomings of early twentieth-century medicine and rescue mechanisms. Unfortunately, even after the first half of the century, developed countries were still ill-equipped to deal with severe shark-attack injuries. Today, powerful intravenous antibiotics, especially helpful against saltwater gram negative bacteria (e.g., Vibrio), have been invaluable in controlling infection, particularly when deep penetrating wounds are inflicted by sharks with spikelike teeth. Carcharhinid sharks and other powerful sharks, which have continuously aligned cutting teeth and great crushing power, are capable of effecting complete limb amputations, slicing away flesh with such efficiency that massive hemorrhage from the cleanly severed skin and muscle is frequently fatal. Successful treatment of such wounds requires enormous amounts of blood and plasma if the victim is to survive.

Few attacks involve numerous bites (which refutes the contention that sharks are stimulated to feed by human blood), and about 25 percent of attacks result in death, with the most frequent cause of fatality being shock combined with severe blood loss. The hands, arms, legs, and feet are most commonly struck, while the head and torso are less frequently involved. Leg injuries that involve the severing of both femoral arteries seem invariably fatal within minutes due to catastrophic blood loss. The compromise of one femoral artery high up toward the groin area usually results in death as well. Often, a rescuer can stop the bleeding from large, severed vessels by firmly pressing anything handy (swimsuit, towel, hand) directly on the wound. This pressure should start while the victim is *still in the water*. Such pressure usually causes the vessel to clamp down in spasm, and clots begin to form. In order to facilitate direct pressure to the wound while the victim is still in the water, a makeshift tourniquet must be applied immediately. Tying a surfboard leash (shock cord) or a dive mask strap around a massively bleeding limb could save a life. Always remember that a tourniquet left on too long (more than fifteen minutes) without controlled release can cause permanent injury as well. If a victim is still alive by the time he reaches shore, the greatest threat is the onset of shock. In South Africa, experience with gruesome attacks has proved that a rapid transport to the hospital has probably killed more victims from irreversible deep shock than it has saved. Many South African beaches are now equipped with first aid packs called shark-attack packs, which contain oxygen, saline, plasma, and morphine for an initial thirty-minute, on-the-spot treatment. Beulah Davis, director of the NSB, in collaboration with two physicians, has published detailed first-aid treatment for attack victims: As the victim is being transported out of the water and placed on the beach, he should be in a sloping head-downward position to increase blood flow to the brain to combat shock. Control the bleeding immediately (start while still in the water) by using pressure points, tourniquets, and elevation of an injured limb above the level of the heart. Leaving tight wetsuit leggings on may actually help stabilize a dropping blood pressure. Notify the Emergency Medical Service immediately and obtain vital signs if possible. Do not give the victim warm or alcoholic drinks. Wrap him or her with a blanket to minimize heat loss. Since movement can aggravate shock, do not move the victim until shock is controlled and a doctor is present. Untrained people should not attempt to help the victim in any way because more harm than good can result from well-meant but incorrect attempts to render aid.

Before we journey back to the Jersey Shore during the unforgettable summer of 1916, here are a few more shark-attack facts and tips for the trip:

The NSB suggests the following to reduce the chances of attack in KwaZulu-Natal:

Swim at netted beaches, whenever possible.

Avoid swimming with an open wound (or when menstruating) as sharks can detect blood and other body fluids.

Don't swim at dawn, dusk, or at night when sharks are most active (and have a competitive sensory advantage).

Avoid swimming in the vicinity of flooding rivers.

Don't swim alone.

When visiting an unfamiliar area, seek local advice.

Be cautious, especially when spearfishing.

The ISAF adds the following:

Do not wander too far from shore since this isolates an individual and places one far from assistance.

Wearing shiny jewelry is discouraged because the reflected light resembles the sheen of fish scales.

Avoid waters with known effluents or sewage and those being used by sport or commercial fisherman. Especially avoid areas where fish are acting strangely or where there are frenzied bait fish or feeding activity is present as with diving sea birds. And be alert for unusual movements in the water.

Sighting of dolphins or porpoises does not indicate the absence of sharks since they all eat similar prey.

Use extra caution when waters are murky and avoid uneven tanning and bright-colored clothing since sharks see contrast particularly well.

Refrain from excessive splashing, and do not allow pets in the water with you because of their erratic movements.

Exercise caution when occupying the area between sandbars and steep drop-offs.

If in shallow water but a good distance from shore, and you see a shark in your vicinity, stand still. Sharks can mistake moving hands or feet for fish.

Display of sharks in the American Museum of Natural History (1916).

CHAPTER 6

Science Stalks the Phantom

The frustrating and demoralizing climate of the summer of 1916 was evident when noted scientists assembled in New York to confer on possible solutions to halt the infantile paralysis plague. At the same time, the convention hall in Asbury Park saw governmental officials assembling to discuss a solution to the New Jersey coast shark plague. Science and the logic-driven occupants of public office unanimously agreed that these were new problems that must be faced head-on. However, it was not particularly scientific for people like Governor Fielder (he called the crisis "evil") to add a demonic label to the shark(s), and it was not very helpful for political cartoonists to paint the shark as everything that was ugly or corrupt about society. One newspaper caricature even showed a beleaguered male swimmer being approached by five sharks each of which represented a vice: whiskey, wasted opportunities, gambling, late hours, procrastination.

The presence of Dr. John T. Nichols at the creek side in Matawan on Friday, July 14, held several points of very profound and even ironic significance. This was a man who, along with two other very respected colleagues, had stated only months before that sharks do not attack living human beings in an unprovoked fashion. His expressed interest and the interest of his American Museum revealed that they were now forced to rethink their original stance. Although Dr. Robert Murphy was an established zoologist and had a keen interest in the New Jersey shark problem, his strength was in the field of ornithology. Dr. Frederic Lucas's forte revolved around the subjects of fossils, taxidermy, and museum displays. Since Nichols was the ichthyologist, he was the obvious choice to venture to Matawan. Nichols, Lucas, and Murphy could have been jointly curious to determine if some new mutation of the shark, or some strange aggressive shark like species, could be present on the New Jersey coast and now opportunely confined in the Matawan Creek. At the least, with all of the fictional and sensationalized news reporting, Nichols wanted to make sure that persons by the name of Lester Stillwell and Stanley Fisher even existed, not to mention confirm that they were attacked by a shark. I like to think of Nichols's trip to Matawan as the move of a true scientist. I believe that Nichols began to reevaluate a previous theory, basing his reappraisal on new data that was becoming available.

After the first two tragedies at Beach Haven and Spring Lake, neither the public nor the scientific analysts seriously considered the possibility of additional attacks. Explaining why the two attacks had occurred seemed a tall enough order. Recall that the U.S. Commissioner of Fisheries, Hugh M. Smith, for example, simply asserted that bathers need have no fear of sharks since dangerous sharks never come close to shore. Smith also explained the freak incident in Beach Haven to be the result of a low tide that marooned a hunger-stricken shark that mistook Charles Vansant for a tasty-looking dog. Even after the reality of the second attack, a seasoned fisherman was quoted as saying that such attacks could not happen again "for a thousand years," not to mention six days.

Subsequent to the events in Matawan, theories, expert advice, and attempts at solving the problem were far from scarce. Dr. Lucas, for instance, suggested that 1916 was simply a "shark year," just as there are times when butterflies, moths, or army worms are overly abundant. There were also the claims that heavy bombing in the North Sea had driven dangerous European sharks toward the American side of the Atlantic. The grisly contention that sharks began to change their dietary habits to include human flesh sprang forth after statisticians realized that some fifteen thousand sailors had lost their lives at sea over the two previous years. Two other theories pointed to hunger as the cause of the sharks' inshore migration and threatening temperament. One theory suggested that the unrestricted German U-boat warfare had restricted shipping of passenger liners, merchant vessels, and cargo steamers, thereby depleting the traditional and expected amount of edible refuse thrown overboard. The other theory suggested that the overfishing of menhaden (moss bunker) may have created a hungry state. Moss bunker are fish that come in such enormous abundance that their harmless schools literally boil the surface of the water. Today, their oily nature relegates their use to cat food, omega 3 fish-oil supplements, and fishing chum, and they are adored by the migrating bottle-nose dolphin (*Tursiops truncatus*). The July 29, 1916, issue of *Scientific American* published an article entitled "Sharks, Man-Eating and Otherwise: The Present Status of a Very Old Subject of Controversy." The author not only conceded that sharks attack but stated, "They are like the hyenas of the sea. And like the hyena, a shark, when he is hungry, will attack living animals; the hungrier he is the less discriminating. A well-fed shark will not attack a man; but any shark physically capable of attacking a human with the prospect of success might get hungry enough to do so."

Beach fishermen in 1916 also noticed, as Captain Cottrell did on his unforgettable walk across Matawan's trolley drawbridge, an unusual abundance of stormy petrels, also known as Mother Carey's chickens. The stormy petrel is an offshore seabird, and the locals wondered whether their presence marked a shift in the Gulf Stream pattern or some dramatic atypical ocean drift. The altered distribution of such water flow, they surmised, would also explain the migration of pelagic (offshore) sharks inshore.

While some tossed around theories to explain the New Jersey fatalities, most of the interested parties participated in the hunt for the culprit(s). A group of Asbury Park Fishing Club members assembled at a tiny inlet in West End, near Long Branch, and devoted an entire night to the shark chase. At the mouth of Lake Takanassee, these men caught a 6.5-foot, 143-pound shark after a ninety-three minute fight and dubbed it the "Takanassee tiger."

Shark bounties were the main motivation for New Jersey fishermen who continued to dispatch one monster after another. However, only the sharks whose stomach contents revealed human remains would garner the man-eater bounty prize. Other enterprising individuals saw profit in capturing any large shark and placing it on display around the state. As the late Johnson Cartan told me, "You could go to the Trenton Fair and see a shark on display. At that time everybody was interested in seeing sharks." In Asbury Park, on July 18, the *Asbury Park Press* reported: YOUR CHANCE TO SEE THE GREAT MAN-EATING SHARK AND SEA MONSTER WEIGHING 1,000 POUNDS: NOW ON EXHIBITION AT THE ST. JAMES HOTEL. The description of that shark sounded like a sand tiger shark, one of the larger somewhat dangerous species off New Jersey.

The trend to display and exploit the shark as a monster certainly has parallels to other periods or places when monsterlike creatures become engrained into local or national lore. Bigfoot, the abominable snowman, and the Loch Ness monster are but a few examples of some natural occurrence being transformed into a sensational monster legend. New Jersey was no stranger to monster myths, and since colonial times, the Garden State had been home to the mythical "Jersey Devil," a winged beast that lurked in the backwoods and the Pinelands. The "Jersey Devil" always seems to appear during those freak periods when observers are unprepared to document it. It appears to individuals who are already frightened by the darkness of a wooded area, or when a party is without a camera, or, even more commonly, when no one in a group wants to put down his beer to take the picture. The "Devil" is such an ingrained and established legend that it even made its way to an *X-Files* episode, and it proudly graces the ice of the state's successful hockey team. The Jersey man-eater, however, was much more than a colonial legend or local lore. It was killing people.

Some truly bizarre reports of sharks being beaten away from shore began to appear and seemed to get worse as the summer went along. One such account, headlined like a baseball score, declared: WHALES DEFEAT SHARKS. The article told the story of a vicious battle between a school of hammerhead sharks off Seabright, New Jersey, and a school of small whales. The sharks were said to have suffered "heavy casualties in the defeat," and the action was said to have been seen by two hundred people aboard the fishing boat *Cape Cod*.

Perhaps even more bizarre than the "battle of whales versus sharks" was the July 15 report that a "horse mackerel" measuring eight feet long and weighing 675 pounds was caught off Seabright in the pound net of Captain John Webber. The struggle with the beast took fifty minutes and the efforts of six men. A sea captain of thirty years' experience explained that if a shark does not get its quarry on the first lunge, it will retreat. The mackerel, on the other hand, will continue to fight indefinitely. He believed that the sharks reported to have attacked people were actually man-eating mackerel. I'm not an ichthyologist, but it seems highly unlikely for a mackeral (*Scomber* sp.) to grow to eight feet and 675 pounds. The fisherman quoted here could have been speaking of a tuna or a marlin, but even that possibility is reaching. Another true bony fish that does reach great size is the docile, ribbon-like oarfish (*Regalecus glesne*). It is likely the longest bony fish. In 1885, a twenty-five-foot specimen, weighing six hundred pounds, was captured off Pemaquid Point, Maine. These "King of Herrings," as they are also called, are not known for being aggressive at all. They slink through fish nets and elude capture so well they are rarely seen. The longest of all oarfishes was seen off Asbury Park, New Jersey, by a team of scientists from the Sandy Hook Marine Laboratory on July 18, 1963. That specimen was estimated at fifty feet in length.

In 1916, the fisherman who came up with the most menacing-looking shark would inevitably have the newspaper men running from far and wide. Photographs were at a premium to reproduce in print, but those who were lucky enough to stand next to a shark or reveal themselves as heros of the day were likely to get front-page exposure. Paul Tarnow, a well-known Belford, New Jersey, pound fisherman, captured a large sandbar shark (or bull shark as you will see) in the Keyport Bay during that summer.

At about 3:00 P.M., six days after the creek attacks, Matawan legend Captain Thomas Cottrell was making his way in from a bluefishing trip off Sandy Hook in his boat the *Skud*. He made the trip with his son-in-law, Richard Lee. On this day, Cottrell and Lee reportedly spotted a large dorsal fin swirling just outside the creek. This shark, lurking near the mouth of the Matawan Creek, would burn in the eye of the captain, whose warnings were scorned days before. Cottrell circled the shark and threw a gillnet over it. The shark became entangled, and Lee pulled the thrashing beast to the gunnel and struck it several times with a piece of iron. The shark was seven feet in length, 230 pounds, and had an eight-inch-wide mouth. The shark's upper jaw had one row of teeth and the lower jaw had two rows. The teeth were slender but triangular at the base, serrated, curved inward, and had a sharp point. It was identified as a ground shark, a blue-nose, or diamond-toothed shark (depending on what you read). The blue-nose was a common name for sandbar shark. The term "blue-nose" may, in fact, simply have meant that the shark was of a blue shade, thereby making it a bluish variation of the sandbar. Other sharks that have been known to display blue coloration are the blue shark (*Prionace glauca*), a mako species, a porbeagle, a dusky shark; some color variations of the great white also include dull gray-blue. Bashford Dean, the curator of the Smithsonian Institute, as well as many local fishermen, were actually considering the local sandbar species as the probable culprit in the New Jersey attacks. Since Cottrell's shark was close to the Matawan Creek, John Nichols sent Fred Kessler, an assistant to Nichols at the American Museum, to make an examination of the shark and draw a rendition of the shark for Nichols's analysis. What became of Cottrell's shark after its capture, or even the possible clarification of its origin, may be more interesting than its actual species identity. Cottrell chose to place the impressive fish on display at the fish house atop the Keyport bridge. The shark was packed on ice and placed in a casketlike box. Outside the fish hut, a sign was placed: TERROR OF MATAWAN CREEK, 10 CENTS A LOOK. For days, hundreds of visitors packed the bridge to gain admission. Cottrell considered embalming the shark for better financial longevity, and he immediately took offers from a motion-picture company out of Hackensack. Model T cars from New York lined the narrow shoulder of the road to see "the killer shark that destroyed New Jersey's summer and killed innocent young men." At the same time, a Long Island fisherman also claimed to have caught the "killer shark." He charged a nickel a peak but rented a zinc-lined coffin from an undertaker to "fancy-up" the display. Since Cottrell's shark was the only beast caught close to the creek, it was only fitting that the energetic and resourceful Captain Cottrell be the fisherman to claim Matawan's bounty money. The strict "man-eating" criteria (human remains in the stomach), however, was never enforced in the case of Cottrell's shark.

The excitement, the reward money, and the zealous desire to appease a revenge-minded and confused public could have influenced more than one illegitimate shark catch. In fact, enterprising individuals began to purchase large sharks from the local (or distant) fishermen, take them to a business or a hotel, advertise them as the New Jersey killer, and report the catch location in vague terms. On July 14, a shark was captured in the pound nets of Long Branch and was identified as a blue shark, nine feet long, 325 pounds. The shark was dragged ashore by its tail and, in the process, snapped at the boot of a man standing on the boat and tore off his "oiler" (oilskin coat). The

Hennessy Fishery crew in North Long Branch was responsible for the fish and quickly sold it to the Long Branch Hotel.

Bill Burlew, the nephew of Fisher's buddy, Red Burlew, was working at Cook's Fishery in Monmouth Beach (just above North Long Branch) during the summer of 1916. During an interview in the 1980s, at his home in Keyport, I clearly and gratefully recall that Mr. Burlew told me he was convinced that the 1916 story was the "true *Jaws*" and that my work would be the ultimate inspiration for the next hit film on the subject of shark attacks. He explained that he had been in the vicinity of the Joseph Dunn attack in Cliffwood on that fateful July day. A few days after the attack, Burlew vividly remembered seeing John Cottrell, a nephew of Captain Cottrell, purchase a large shark at the Cook's Fishery in Monmouth for $5. John Cottrell boasted to Burlew that he was going to take the shark back to Matawan and place it on display. Mr. Burlew, however, was not certain whether this was the same shark purported to be the "Terror of Matawan Creek."

The speculation over John Cottrell and the *Skud*-landed shark may have a bizarre connection to the shark that was battered by an oar off Asbury Park during the much-publicized Benjamin Everingham life-guarding incident. A dead shark was found near an outflow pipe by an engineer in Monmouth Beach on July 16. The shark was a blue-nose shark, weighing several hundred pounds, measuring seven feet long, and with an eight- or nine-inch-wide jaw. These were the same dimensions as Cottrell's bounty-winning shark, and it was found in the town in which Bill Burlew claims he saw John Cottrell purchase a large shark. The Monmouth Beach shark was said to have been bleeding from blows attained from the oar in Asbury. The story behind the capture of Cottrell's shark becomes even more suspicious when one considers that the shark which was displayed on the Matawan bridge was said to have been caught in a gillnet and "clubbed with a piece of iron several times." What better way to explain the shark's imperfections (oar-induced injuries?) but by the claims that it was struck by an iron club during the frantic catch.

Besides the eyewitness testimony on John Cottrell's purchase, Bill Burlew also informed me that he recalled seeing more sharks in the Keyport Bay in 1915, 1916, and in 1917 than in any other years he could remember. Burlew's comments are fully consistent with the other reported shark sightings in the years he mentioned.

The evidence that many sharks were migrating shoreward in 1916 could have great significance in determining whether one shark or several sharks were responsible for the entire summertime horror. Edwin Thorne of the Board of Managers of the New York Zoological Society corroborated Burlew's testimony by documenting his catch statistics for the summer of 1916. Thorne reported that he captured or sited more female brown sharks (another name for the sandbar shark) that summer in the Great South Bay of Long Island than in any other season of his seventeen-year study. Six of the sharks killed in those years were dusky sharks. Between 1911 and 1927, Thorne sighted 1,799 sharks and killed 305 of them. In 1916 alone, he saw 277 sharks and killed 102. The sharks were apparently in the Great South Bay to give birth to young. Thorne wrote:

> In the past season [1916], my man on the look-out at the mast-head saw 82 sharks during one forenoon, and I counted 42 from the deck at the same time. They were of course unusually plentiful on this particular day, although I believe that two hundred during one entire day is a low estimate of the number seen on occasions before I began to keep an accurate record.

In August 1916, Robert Murphy joined Thorne on an excursion. Murphy wanted to obtain specimens to reproduce life-size models for the museum. Murphy was pursuing an artificial method and casting system for display (which all modern museums now use). Dr. Murphy simply could not accept the stuffed versions at other exhibits, which inevitably ended up with cracked skins, warped scales, and shrunken fins. On their journey toward the north end of the Fire Island Lighthouse, the men were barely two miles into the Great South Bay when the harpooner was called into action. He steadied himself, and at the favorable instant, he drove the iron deep into the shark's body. With a rush, the stricken creature sped away, carrying yard after yard of rope, "the end of which was fastened to a tub. . . ." The men caught a large sandbar shark (six feet) on that try and also caught a large dusky shark (eight feet, 332 pounds) that afternoon. Even though Thorne's study lasted seventeen years, the seasonal weather restrictions limited the outings to just 302 days in the field. Those numbers give an average of 106 shark sightings per year. As mentioned, in 1916 Thorne recorded 277 shark sightings and 102 bagged. In the end, Thorne's findings do seem to support Dr. Lucas's assessment that 1916 was a "shark year."

One other influential factor in the 1916 attacks also supports the observations of Thorne. The New Jersey pound nets were a type of catch-net apparatus and temporary farm "cage" for fish that were netted or otherwise caught by commercial fishermen. The pound nets were placed within a mile of shore, and they were prevalent along the Jersey Shore from the 1800s through the 1930s. The catching action of the nets was apparently executed by the rising and falling of the tide. Their very presence may have been an influence in luring large sharks inshore. The problem with this theory, however, is that pound nets were present for decades without one attack ever being documented.

That pound nets were in place for quite some time without the occurrence of shark attacks doesn't mean that they were not responsible for influencing other strange events. Ned Ralston, an old salt from Allenhurst, told me a story of a yellow fin tuna he and his partner, Malcomb Carton, caught just around the pound nets in the late 1930s. They were coming in from a long-distance fishing trip and decided to drop their lines near shore. The tuna was close to a thousand pounds. In 1916, fishermen reported finding sharks "five and six more times" than usual in the pound nets.

Edwin Fowler, a specialist on New Jersey sharks during the 1916 period, also appraised the photos and species of sharks being caught and felt that the vast majority of the sharks were sandbar sharks. The sandbar shark has the habit of appearing as it crosses a sandbar, then disappearing again on the other side. Fowler even went so far as to attribute the New Jersey attacks to such a species. The sandbar is also called the New York ground shark because, early in the century (and in 1916) hundreds were seen at a time in Great South Bay, Long Island, between Lindenhurst and Great River. Hundreds of sharks were undoubtedly caught in the summer of 1916 in New Jersey, and many more were sighted. The sandbar was obviously one of the more prevalent sharks caught, but photographs and descriptions confirm that a great many other species were snared as well. As mentioned, the description of a shark placed on display on July 18 in Asbury Park sounded much like a sand tiger shark. Near the Matawan Creek that summer, a newspaper description of sharks that had been spotted sounded strangely like scalloped hammerhead sharks (*Sphyrna lewini*). Although the witnesses I spoke with never supported the assertions that the creek was home to several sharks, the newspaper accounts spoke of the Matawan Creek being alive with many small sharks and that the Wyckoff dock area was a spawning ground. Other reports spoke of three sharks being cornered at one time in the creek, only days after the attacks.

The presence and capture of large sharks so close to New Jersey's coastline was certainly an anomalous circumstance. Most of these sharks were caught, hung upside down, placed on display, then discarded. Few were sincerely touted as "man-eating species" by their captors, but for purposes that were pleasing to the public, they were all "man-eaters." As for their stomach contents, only the false Keyport report touted fictitious human remains. Very few, if any, reported even suspicious (human) digestive remains. There was one report about a woman's foot and tan slipper being discovered in a shark off Spring Lake in 1914, but as for 1916, only one shark, as we will see, did have documented human remains in its stomach.

At the American Museum of Natural History, Dr. Lucas was sifting through the data that Dr. Nichols had compiled during his trip to Matawan. Nichols could not help but tell Lucas how he was enamored by the small midwestern-like town in New Jersey. Nichols felt as if a strange tiger from the sea had invaded the sanctity of a peaceful and undisturbed setting and turned it upside down.

Dr. Lucas, in reply to H. F. Moore, Acting Commissioner of the Federal Fisheries Bureau, wrote: "I hasten to reply to your letter of July fourteenth, in order that you may know we shall be very glad indeed to send you any information that Dr. Nichols may obtain." Lucas closed the letter with the humbling recognition that sharks were the cause of the "shark plague." He wrote, "There seems to be no question that the fatalities which have occurred are due to sharks, but the species is still undetermined."

Even beyond having Lucas acknowledge shark involvement, there was something more about Nichols's Matawan excursion that spurred him to become even more motivated to get to the bottom of such a tragic and unusual situation. Possibly, after seeing the sincerity and sentimentality in the eyes of Matawan residents, Nichols became obsessed with solving and closing the case.

Nichols informed Lucas that he and Robert Murphy would undertake a daylong expedition to hunt down the northward-moving man-eater. Nichols was now virtually certain that a single shark was the true killer. Lucas, by using the word "species" in a singular context in his letter to Moore, must have also felt that a single *type* of shark was involved in the attacks. As mentioned, Nichols's theory related to the likelihood of an exotic species causing the damage. In a *New York American* article, Nichols stated:

The most plausible theory seems to me that the trouble is from one or several exotic sharks which are working up the coast; and we have determined to spend a day or two looking for it off Rockaway, in case it should cross to that side of the harbor, as is not unlikely if our surmise is correct. Unless the shark came through the Harbor and went through the north through Hell Gate and Long Island Sound, it was presumed it would swim along the South Shore of Long Island and the first deep water inlet it reaches will be the Jamaica Bay.

Nichols and Murphy were going to lay in wait at the narrowest part of the channel. The men considered a tiger shark, or, more likely, a white shark that had strayed three hundred miles from its natural habitat, to be at work. If the white shark were spotted, it would be only the second time in history that one had been present above Cape Hatteras. "The garbage in New York Bay and the chances of catching unsuspecting swimmers will undoubtedly bring the sea tiger to New York's waters," Nichols stated. Nichols did not comment on why he thought the shark was present, but *Scientific American* later equated the matter to a man taking a different street on his way to work.

The mission that Nichols and Murphy undertook to capture the northward-moving monster of the deep may seem like a task made to order for portrayal on the silver screen, in fact, for portrayal by the likes of Robert Shaw, Richard Dreyfuss, and Roy Scheider. Nichols and Murphy did not capture a man-eater in their attempt, but their failure should not paint the attempt as far-fetched or foolish. Such an attempt may have even gotten them the culprit they were seeking if it had not been for yet another most peculiar event. The culprit they were seeking may have actually been caught the day before.

On the very same morning that little Lester floated to the surface of Matawan Creek and the morning that Stanley Fisher's family arrived back in town, a catch was made in the Raritan Bay that drew the appraisal of a great many eyes, of both scientists and laymen. During the very hour that Dr. Nichols was investigating the scene at Matawan Creek, New Yorkers Michael Schleisser and John Murphy set out for Raritan Bay from South Amboy, New Jersey. Schleisser and Murphy were almost as far south as the Sandy Hook Bay when they dropped a dragnet at the stern in a casual effort to catch some pan fish for breakfast. After an hour or so, the men were far from their place of departure and only a few miles from the mouth of the Matawan Creek. Their net was about six feet under the sputtering craft when, without warning, a tremendous tug and jerk lurched the boat. The abrupt halt stalled the engine, and Schleisser, near the stern, spotted a big black tail fin and shouted, "My God, we've got a shark!" The tiny vessel was lunging backward at great speed, and the bow was riding dangerously high out of the water. Waves were crashing over the transom. Murphy cleverly and desperately threw himself onto the bow to prevent the stern from going under. The shark was now thrashing so violently that the entire stern area was white foam, and Schleisser was in the midst of it. Schleisser couldn't help but notice that the shark was as big as his boat. His sense of dread must have been pitching at a rate that shot off the scale.

Sometimes, however, cool-headed desperation works with luck. Schleisser suddenly remembered that he had thrown a broken oar handle into the boat at the dockside when they left South Amboy. Schleisser, a Barnum & Bailey lion tamer and animal trainer by trade, began to strike the shark's head, nose, and gills with every ounce of force that his modest frame could muster. The more the shark thrashed, the more it became entangled, and the more Schleisser was able to administer deadly blows.

After he was certain that the once-crazed shark was incapable of producing so much as a twitch, Schleisser and Murphy signaled a larger boat in the vicinity to help them with their most unusual morning haul. The crew of the approaching craft knew immediately that the vicious net fight and the stern-first "Nantucket sleigh-ride" was not a fish story. The shark was harnessed to the side of the larger boat and towed a few miles to a dock in Atlantic Highlands, then back to South Amboy.

The fishermen who assisted the new Friday morning heroes were quite interested in the shark. They recognized immediately that this shark did not look like the plethora of others being caught. One look at the front end of this streamlined, muscular specimen, and the growing crowd knew that this monster had only meant business.

Whenever a large shark was captured during this period, the lucky fisherman was faced with a short-lived dilemma. Should he cut the shark's digestive tract open to determine if any unusual and potentially bounty-winning dietary items were inside, or should he leave the shark's mighty skin in pristine shape for the benefit of display aesthetics? Some fishermen were content with inverting the shark (hanging it upside down and watching for dietary items that might drop out of the mouth) and assessing the stomach contents in that way. Schleisser's beast, however, was towed to shore and not suspended. Additionally, Schleisser and his partner, as well as the gathering spectators, felt something eerie in the air about the entire event. With all of the shark hunters sporting explosives, shotguns, heavy-duty poles, and handsome portions of beef and lamb, how could it be that an unprepared duo could venture out into a bay, in a miniscule motorboat, without even a fishing pole, and return with a terrifying man-eater? This was an episode as bizarre as the 1916 attacks themselves.

With all of the compelling reasons to make the move to dissect and thrust a shark knife into the flesh of the Raritan Bay shark, perhaps the biggest one was the fact that Schleisser was a seasoned, if not renowned, taxidermist. His specimens still adorn the display cases throughout the most prestigious museums of the Northeast. It was Schleisser, therefore, who firmly grasped the knife and plunged the tip of the sharp blade just inferior to the hard cartilaginous chest region. He skillfully slid the weapon in and down. Only the grown men in the crowd could peer over his head to visualize the cutting. When pungent yellow-brown intestinal juices began to flow along the dock boards, only the hardiest of spectators remained near the action.

Schleisser spread the tough underbelly to expose the vulnerable viscera. To the crowd, Scheisser's cool demeanor made it appear as if he'd done this very operation hundreds of times before. The huge, brown, smooth bi-lobed liver slid out first. Then, with some gentle coaxing, the bulbous stomach and the shear-skinned tubelike irregular intestine wriggled into view. Schleisser felt at the thicker stomach cavity and found it difficult to determine the contour of its contents. With the intestines, however, one grasp with two hands and he, as well as the rest of the onlookers, could distinguish assorted items that appeared firm, narrow, and suspicious. Just as Schleisser was about to make the cut that would finally reveal the identity of this most interesting diet, a local dentist observing the operation said he thought it was important to have a medical person in attendance. The dentist knew that Schleisser was seconds away from slicing through digestive organ linings and exposing, he hoped, the food items within the shark. Those items, the dentist realized, could range from easily identifiable marine creatures to, in this case, body parts from a recently killed boy. Some of the spectators may have known about the sensational *Star-Eagle* headline that had come out that very morning: HUMAN BODY FOUND IN SHARK. And they may have even still believed the headline. One should remember that television and radio was not available to make quick corrections. If members of the crowd were under the false impression that Lester Stillwell's body had already been recovered in the shark caught in Keyport, they might have had some other misconceptions about identifying the shark's stomach contents. This would mean the culprit at this dock was probably not the Matawan Creek shark and could only contain remnants of the oceanfront victims (i.e., Bruder or Vansant). As you will see, the examiners call it the way they see it, regardless of what the newspapers said about a "boy's body."

When the level-headed dentist returned, he brought along two friends, physicians who were just about ready to head to their local offices. On the dock, much of the crowd had now dissipated because they had not seen the clearly identifiable human remains they'd hoped for. Around the slick of bubbling digestive acids lay portions of other unidentifiable fragments. Schleisser and Murphy had separated out a portion of suspicious fleshy material and bones that took up "about two-thirds of a milk crate" and together weighed fifteen pounds. Among the bones were small fragments, a long straight bone, and another smaller narrow portion. The physicians at the dock identified themselves to the men in charge and immediately keyed in on the fleshy tissue. They and Scheisser knew that the surface flesh was not that of a shark, a ray, or of standard fish origin. The men concluded that the pale and partially decayed outer surface tissue was human epidermis, and the underlying fat and fascia were also consistent with deeper human tissue layers. The straight bone had one narrow, chipped end and one flared end. It was determined to be the eleven-inch shinbone of a boy. The smaller piece of bone was identified as a human rib portion.

Schleisser was not fully surprised by the assessment made by the dockside physicians, and, being ever resourceful, he prepared to haul the beast and its suspicious contents back to New York in his motorcar. Schleisser's plan was to get the shark back to the Bronx and prepare it for preservation. He knew it would have been difficult to transport the carcass fully weighted by intestines, stomach, and other internal organs, and he also knew that the heat would do a number on the perishable innards. He presumed that the shark, gutted and still thick-skinned, would make the hour-long trip without incident. As for the larger bones, he placed them in alcohol and thought they would be just fine for the trip. He was not optimistic that the few pieces of chipped ice available would prolong the state of the human flesh. Schleisser and his unorthodox cargo did make it back to the Bronx in one piece. Unfortunately, the flesh recovered was discarded by Schleisser as its "smell became quite offensive."

As a career taxidermist, Schleisser was a man who had frequent contact with museum heads and assorted scientific figures. He knew that the discovery of potential human remains in the stomach of his shark obligated him to seek consultation from the appropriate authorities, and he knew that those authorities resided at the American Museum. As we will see, the accepted and conclusive identification of the stomach contents of the Raritan Bay shark was far from finalized at the dock.

On Saturday, July 15, in a multirow specimen storage room at the American Museum, Dr. Nichols returned with Dr. Murphy from a late and frustrating day of trying to capture the New Jersey marauder. Nichols was in the middle of examining some shark teeth when an exasperated Frederic Lucas walked in. Lucas spoke of yet another tragedy that had taken place a day before. Lucas explained that a man from Newark, Samuel Harding, thirty-five, had been swimming off Atlantic Highlands, New Jersey, in the Shrewsbury River. Harding threw up his hands, called for help, and was in obvious distress in front of the many swimmers and spectators. Just as two or three strong swimmers were about to dash to his aid, someone yelled, "Shark! A shark's got him." About twenty minutes later Harding's body was retrieved by an oysterman's rake and didn't have a scratch on it. "They let the poor man drown," Lucas shouted, "just because they thought he was being eaten by a shark." In the case of this unfortunate man, there was no shark at all. Perhaps he should be considered yet a sixth victim of the Jersey man-eater.

As Lucas and Nichols were in the midst of conferring over their next move, a messenger from the Brooklyn Museum rushed in with a newspaper article sent by Dr. Murphy. Dr. Lucas snatched the paper, the *Bronx Home News*, from the messenger and read the front-page headline aloud: "TWO MEN IN TINY BOAT CATCH KILLER SHARK, BEAT IT TO DEATH AFTER SHARK TANGLES ITSELF IN DRIFT NET. . . . PART OF BOY'S SHIN TAKEN FROM LARGE FISH CAUGHT NEAR CITY." Nichols leaned over Lucas's shoulder and spotted a paragraph that mentioned that "physicians on the dock report that the shark contained fifteen pounds of human remains, including a boy's shinbone and a rib section."

"Perhaps this is something we should look into?" he asked.

Lucas read on that the shark was being prepared for display at the *Bronx Home News* office, and although he was still skeptical about the current reports, he suggested that they consider a trip to see this particular specimen.

The messenger had barely left the room when an elderly secretary entered to hand-deliver a parcel to Lucas. The package was from Michael Schleisser and was labeled OPEN AT ONCE. The parcel contained the bones scavenged from the stomach of the Raritan Bay shark. The men decided to go to Lucas's office. Nichols tossed the newspaper on the windowsill. The headline below the fold read: U-BOAT *DEUTSCHLAND* FAILS TO ARRIVE AT LONG ISLAND PORT, FEARED SUNK OR LOST.

In the small box of bones, Schleisser enclosed a letter describing the shark as "dark-dull blue, with a white belly, and the mouth, when open, can fit a man's head inside. It has four rows of teeth. It is 7 1/2 feet in length and weighed 350 pounds." Other newspaper reports from this period cite the shark as being 8.5 feet in length. Schleisser also mentioned that he was working at a torrid pace to prepare the fish for display at the main window of the *Bronx Home News,* scheduled for July 19 and 20.

Lucas immediatcly responded with a follow-up request about this increasingly interesting catch. Lucas wrote to Schleisser: "I am very much obliged to you for your courtesy in letting me see the bones taken from the shark. They are the parts of the left radius and ulna [lower arm bones] and one of the anterior left ribs, all human. There is no doubt about this. They have been badly shattered. Can you tell me the exact species of shark from which these were taken? Or, if you are in doubt, I'm sure Dr. Nichols would be very glad to call and determine the species exactly."

Lucas's letter to Schleisser underscores each scientist's strength. Nichols, for example, was the man to look to for the identification of the shark. He was a Harvard educated ichthyologist who had held positions with the American Museum of Natural History and the U.S. Bureau of Fisheries since 1908. His expertise was in the field of marine fishes, and before the end of his career he would author two books on fishes and produce some 950 scientific and popular articles. Likewise, Dr. Lucas was fully qualified to assess the bony remnants before him. Lucas's illustrious scientific career began in 1871 with his position at the prestigious Natural Science Establishment at Rochester, New York. There, Lucas specialized in the preparation and mounting of skeletons. He developed such a high technique that, in 1882, he was called to the U.S. National Museum in Washington as an osteologist (bone expert); by 1902 he was simultaneously the curator of comparative anatomy and fossil vertebrates.

Obviously, Lucas's anatomical assessment of the items transferred to his office was quite a dramatic turning point. Not only was Lucas interested in learning more about the shark from which they were taken, but he, as a noted anatomist, although not in strict agreement with the dock physicians regarding bone identity/anatomical location, confirmed that they were, in fact, human.

On July 15, at St. Peter's Hospital, the sister superior, who was now acting as hospital spokesperson, stated that Joseph Dunn would survive if infection did not set in. She also reported that the doctors might even save his leg. On the first day that Joseph was admitted, he refused to tell any officials where to find his mother for fear of worrying her. She had been contacted without his knowledge and was told that her son had been almost swallowed alive by a shark. The next morning he woke to find her standing at his bedside. The first words he whispered to her were, "I'm all right, Mother; it didn't hurt me a bit." Joseph's roommate, Frederick Hunt, was interviewed by the press and couldn't believe how bravely Joseph was taking the whole ordeal. Hunt had recently been injured in a factory-belt accident. He mentioned that he had thought his injuries were bad until he looked over at poor Joseph. Hunt now felt fortunate and mentioned that he doubted he could ever tolerate the shock that Joseph must be enduring.

During the first forty-eight hours of Joseph's hospital admission, he was weak and groggy from loss of blood and the aftereffects of ether used during surgery. During the evening of July 13, Joseph was unable to get a good night's sleep because of a thunderstorm—the same foul weather that supposedly drew Lester Stillwell's corpse to the surface of Matawan Creek. At 10:30 P.M. on July 16, at the pediatric surgical ward at St. Peter's Hospital in New Brunswick, Joseph Dunn tossed and turned and intermittently shouted, "Help! help!" Joseph was now becoming more alert, and medical staff saw his arms were moving frantically under the covers. His one good leg was also seen struggling against his hospital sheets, as if he were fighting a phantom. Hospital staff reported that he tore bed quilts and pillows and shrieked in terror. Joseph was having a shark-filled nightmare. It had been four days since the once-in-a-lifetime traumatic event, and this lone survivor of the Jersey man-eater was still fighting off the idea of being swallowed whole. He was far from walking, but his lower left leg was still viable. The early and ugly skin grafts around his calf were constantly monitored by his doctors, and the nurses were doing everything they could to keep the site moist, clean, and infection-free. The skin donor sites at his lateral thigh regions were red, sore, and weeping.

Joseph's will to live was surpassed only by his brave spirit and his pleasant personality. The hospital staff dubbed him "little Jonah" after the regurgitated biblical prophet, and his posttraumatic stress was eased on a daily basis as he held court with a host of ward children who'd flock to hear his tale of the man-eater. Just that morning, Joseph had sat up and read dozens of get-well cards from around the country. In between letters, he would speak of how he faced the shark head-on and got some help from his brother and their friend Jerry.

After the capture of the Raritan Bay white shark, noted scientists presented fantastic appraisals of shark ferocity, and some revisited the Jonah legend with the insertion of the white shark as the whale/great fish. Dr. Hugh M. Smith, considered by many in 1916 to be the foremost American authority on fishes and who in early July said that "bathers need have no fear of sharks," was now making up for his past statement and was publicizing a new communication to the National Geographic Society. In a late July *Newark Star-Eagle* article entitled A SHARK MAY HAVE SWALLOWED JONAH: SCIENTIST SHOWS BIBLICAL "GREAT FISH" HAS COUNTERPART IN PRESENT-DAY MONSTER, Smith noted that some gigantic shark specimens are as "harmless as doves and others the incarnation of ferocity." Smith goes on to state, "One of the most prodigious, and perhaps the most formidable of sharks is the man-eater, *Carcharodon carcharias*. It roams through all temperate and tropical seas, and everywhere is an object of dread. Its maximum length is forty feet and its teeth are three inches long." Smith finally proposes his explanation as to why, before the summer of 1916, there were few authenticated instances of shark attacks on human beings. Apparently Smith believed that earlier victims were simply not accounted for. He states, "There have undoubtedly been many cases where sharks simply swallowed their victims whole as with the story of Jonah."

Robert Murphy, originally one of the staunchest skeptics of a shark's capacity to attack a live human and sever a human bone, was now not only convinced that sharks do attack humans, but also that the Raritan Bay white was the guilty specimen in the New Jersey killings and the same type of shark that swallowed Jonah. In an article in which Murphy quotes Linnaeus's early thoughts on the subject, the headline read: FISH THAT ATE JONAH CAUSED DEATHS HERE. In the July 16, 1916, edition of the Sunday *Philadelphia Record,* an interesting story was run on sharks; in the layout they presented a relatively famous reconstruction of the huge jaws of the prehistoric *Carcharodon megalodon.* The teeth of *C. megalodon,* the forty-five-foot ancestor of the white shark, have been found in many locations throughout the United States and

because New Jersey has particularly abundant marl areas (a green natural clay with preservative properties), it has provided some nice specimens of these teeth. In fact, New Jersey was the first location to provide these teeth as well as the first nearly complete dinosaur fossil
(from Haddonfield). *Megalodon* teeth have been found as large as
seven inches long and the reconstructed jaws in this particular article
appear to be the same jaws displayed at the Hall of Fossil Fishes at the
American Museum of Natural History (originally constructed in
1907). In one photo of the jaws, six ichthyologists are situated inside
the jaws for illustrative purposes. In the July Sunday *Record* photo, only
one man is present in the immense jaws but also within the jaws is
printed the statement: "Jonah could have been easily swallowed by this
80 foot prehistoric man-eating shark." Today, paleontologists and
ichthyologists generally agree that the Megalodon grew to between 45
and 50 feet. Incidentally, for the more famous picture of the six
ichthyologists in the jaws (two kneeling and four standing), the caption under that photograph was prepared in 1940, but I believe that
the jaws were actually reconstructed even prior to 1916. I also believe
that the man in the far right corner of the photo (standing) is not an
ichthyologist at all, rather a taxidermist named Michael Schleisser.

On the afternoon of Wednesday, July 19, a week after the Matawan
horror, Michael Schleisser had broken all previous taxidermy speed
records and opened an exhibit of his man-eater at the *Bronx Home
News* building. On this very day, President Wilson was trying not to
think of sharks and instead began daily visits to the summer executive
headquarters at Asbury Park from his residential estate at Shadow
Lawn. The *Home News* office building in the Bronx, however, created a
stir in New York that made the entire town "shark crazy" with excitement. They came in swarms from the Bronx, Washington Heights, and
even from Mount Vernon and Yonkers. The *Home News* touted the
story: HARLEM MAN IN TINY BOAT KILLS A 7 1/2 FOOT MAN-EATING SHARK:
BEATS IT TO DEATH WITH BROKEN OAR, DIRECTLY OFF MATAWAN CREEK. EXAM-

INATION BY DIRECTOR OF MUSEUM OF NATURAL HISTORY SHOWS HUMAN
BONES IN SHARK'S STOMACH!

At the newspaper office, extra policemen were called out to deal
with the crowd of spectators. Some thirty thousand people blocked up
the sidewalk in their efforts to catch a glimpse of the monster they
read so much about. According to the *Home News*, everyone wanted to
see "the yawning jaws and vicious teeth of the sea monster." The trolley line was jammed all day Wednesday, and for two consecutive nights
until 10:00 P.M. the crowds were rows deep in front of the impressive
toothy display. A frequent question from amazed spectators was, "Is it
real?" Others simply questioned its anatomical authenticity and accuracy. One confident onlooker stated, "It is not a shark, a shark's mouth
is on its side." One little girl called it "cute," and her mother even suggested it might be a porpoise.

Late Thursday afternoon, Nichols, Murphy, and Lucas jumped off a
trolley across from the *Home News* office. A line of onlookers wrapped
around the block, and as Nichols and the group attempted to bypass
the exhibit line, they were questioned by a frazzled clerk and a pretty
receptionist. Inside, behind the display window and away from the
mob, the front office was packed with reporters. Playing the role of
ringmaster, Schleisser told the chilling story of the shark capture and
the stern-first ride he and his friend were forced to take. Schleisser
brought along the boat oar and explained that, although he had
tamed and encountered hundreds of ferocious beasts in his travels, he
never had such a dreadful fight for life of this magnitude. The proud
Schleisser confirmed that the shark did indeed possess human remains and boasted that he stuffed the animal carcass in world-record
time. Schleisser also explained how the shark's teeth and open jaw
caught him between the elbow and the hand during his battle with the
man-eater, and even though the beast had no chance to clench down
on his forearm, the skin of the arm had swelled up dreadfully. He even
had bandages over his fist and knuckles where abrasions were made by
the sandpaper-like hide of the monster.

When Schleisser completed his brief presentation, a reporter shouted a simple question: "What kind of shark is it?" Schleisser appeared unprepared for the inquiry and began to describe the shark's dimensions. Lucas, Nichols, and Murphy proceeded a bit closer to the display to inspect the shark more carefully. Within moments, Dr. Nichols informed Schleisser and the crowd that he had captured a young *Carcharodon carcharias,* a great white shark.

The shark experts were halfway out the newspaper office door when they overheard Schleisser announce plans to depart with his collection of finely preserved specimens, as well as the now-famous white shark (the top contender for the Jersey Man-Eater title), on a magnificent tour throughout the Far East. "In two weeks," Schleisser stated, "I plan to depart with my most impressive specimens, and return with some rare animals as well. I am heading on a two-year excursion to Nagasaki, Japan, an unknown region of Tibet, and Central Asia."

A disappointing footnote to Schleisser's two-year Asian excursion is that the white shark, which he so quickly and enthusiastically embalmed, likely did not make it to a prestigious U.S. museum after all. I have been unable to locate any records of the mounted animal's current whereabouts or where it came to rest when it returned to America. The only existing photo is that which appeared in the *Bronx Home News* of that period. I will never forget the day I unearthed the photo in the archives of the *Home News.* My first move was to send a copy of the picture to Richard Ellis, who was working on *Great White Shark.* Even though Ellis's book was already being copy edited in San Francisco, he thought so highly of the image that he sent out a cry of "stop the presses" to allow for the photo's inclusion in that edition.

As for the stuffed version of Schleisser's shark, I fear that the inferior early-twentieth-century methods of preservation, as alluded to by Dr. Murphy, compounded by the fact that this was "rushed" taxidermy, led to a deterioration of the man-eater's carcass. However, the jaws of that specimen may have been extracted and separated from the head of the once intact carcass when Schleisser returned to the United

States. In 1918, the renowned ichthyologist E. W. Gudger walked into a fish shop on Broadway and Eighty-sixth Street in Manhattan and couldn't help but notice an unusual set of jaws decorating the wall of the establishment. Gudger astutely noted the jaws to have an upper and lower set of triangular serrated teeth which were unmistakably from a great white shark. Gudger, who remained with the American Museum of Natural History until the age of eighty-eight, made this statement in a 1950 paper that examined the attack on a boy in Buzzards Bay, Massachusetts. In the paper, Gudger wrote: "I examined these jaws and noted the characteristic broadly triangular saw-edged teeth, which showed that these came from a *Carcharodon carcharias*— and presumably from the New Jersey shark of 1916." The mounted jaws that Gudger spotted were labeled "The Jaws of the New Jersey Man-Eater." Unfortunately, the current home of those jaws is also unknown. Hoping for a miracle, I did do quite a bit of legwork by perusing a multitude of modern fish markets in the vicinity of New York City. Perhaps someone tossed the jaws in the trash decades ago, or perhaps they now grace the living room wall of a local collector.

Besides the mention of the jaws and the description of the attack at Buzzards Bay, Gudger made comparisons to other attacks in that 1950 publication. The paper suggested that the bite marks and attack behavior of a medium-size great white in the Buzzards Bay attack on the boy appeared strikingly similar to the wounds and behavior relating to the attack on Stanley Fisher.

In 1916 and for decades to follow, the lay and scientific communities were without a consensus reason to explain the bizarre happenings of that summer. Several imaginative northeast residents connected the strange occurrences of July 1916 with the concurrent presence of German maneuvers on U.S. soil. The dastardly "cigar bomb" explosions perpetrated by Franz Von Rintelen, otherwise known as the "Dark Invader," were already causing chaos in the northeast U.S. merchant ports, which were sending goods overseas. Although America was not yet in a state of war with Germany, large ports like those of Newark, New York, and Boston were all wary of the possibility of small cigar-shaped bombs being neatly planted in cargo merchandise headed for other shores. Ships would leave port, get a few miles to sea, then suffer a small explosion that would disable it and send it slowly to the bottom. The explosions, believe it or not, were meant to harm only the ship and cause as little human death as possible. The "Dark Invader" was a "thoughtful" terrorist. Von Rintelen fled the United States when authorities became privy to his every move. He was later detained in England or Ireland and deported to Germany. Even more bizarre was a botched espionage mission that took place in 1915. On a Manhattan elevated railway car, one year prior to the shark attacks, a passenger came across a most interesting piece of material misplaced by a German spy. After Heinrich Friedrich Albert, an agent for the German government, carelessly left a briefcase filled with sabotage plans on the railway car, the contents of the case not only foiled the plot, but were traceable to the German and Austrian embassies.

The mysterious-looking German U-boats, which appeared on the U.S. coast in the summer of 1916, also added fertile material for imaginative minds. The *Deutschland*'s captain first raised suspicions about Germany's intentions when he refused an inspection by U.S. officials. At first, the boat was said to be carrying a message to President Wilson from the Kaiser, then the message was said to have been a hoax. Some officials considered detaining the boat until the war was over. On July 12, the Holland government even sent word to the United States that the presence of the German U-boats in American waters was meant as a warning that the United States was in reach of Germany's powerful forces. Some panicked East Coast residents actually considered that German covert activity was the cause behind the shark problem. Even as early as July 8, the *New York Times* wrote: "A reckless imagination might suggest, as an alternative hypothesis, the possibility of Charles Bruder fouling the propeller of the German submarine that is or is not somewhere in these waters, but the theory must be rejected as soon as it is heard."

The German submarines *Bremen* and *Deutschland* were, in fact, cruising the mid-Atlantic coastal states during the general shark-attack period. The July 10 *Washington Post* front-page headline read: GIANT GERMAN SUBMARINE ARRIVES IN CHESAPEAKE CARRYING DYES AND MESSAGE FROM KAISER TO PRESIDENT. It was the *Deutschland* specifically that was touted as the first underwater liner and the world's first and greatest undersea boat. The *Deutschland* was referred to as a "merchantman," but it carried two small-caliber guns and a crew of twenty-nine. Its status as a warship was a hotly contested issue among officials at the State Department. It made the 4,180-mile trip from Bremerhaven to the Virginia Cape in a speedy sixteen days. On July 14, (two days after the last attacks) its commander said it was leaving Baltimore, Maryland, and heading back to Germany. However, after several days unaccounted for, the sub suspiciously reappeared in Bridgeport, Connecticut. Newspaper cartoonists could not let the opportunity of simultaneous visitation by two unwanted entities (the sharks and the subs) pass them by. The comics revealed such caricatures as a big black submarine, replete with front-end teeth, snapping at a vulnerable Uncle Sam. Could the Germans have created a shark-attracting device or an apparatus that altered the temperament of sharks, prompting them to feast on tasty human bathers?

Simply dismissing the possibility that German military scientists of 1916 could have created a shark-attack-stimulating device would probably be the safest and easiest path to take. To dismiss such a theory by assuming that the Germans were not highly advanced in certain fields would be an injustice. In World War II, for example, the Germans were impressively ahead of all other countries with the sophistication of the baffling code-communicating device, the enigma machine. And not only were they first to use semiguided missiles (i.e., the V-2 rocket), they were also experimenting with flying discs (saucers). Even though the semiprimitive subs of World War I were not capable of mustering the successive decisive blows of the more sophisticated underwater crafts of World War II, they were brutally successful at sinking merchant vessels without warning. The Germans wrongly claimed that their submarines were fragile and were easy targets of ships' guns when on the surface.

The very fact that Germans were using submarines so extensively and successfully was a testament to their ability to create and utilize fairly novel methods of mechanization. Many historians believe that the Germans were well on their way to winning World War II with the auspicious success of their disruptive submarine warfare. The United States and its Allies, however, eventually realized that the subs were obligated to spend several hours on the surface of the water to recharge. It was at that time that they were most vulnerable to air attack. It was only the pioneering wartime perfection of radar that made these surface visible subs vulnerable as air strike targets. In a move that changed the tide of the war for the Allies, scientists from Britain, the Massachusetts Institute of Technology, and the Twin Lights lab in Atlantic Highlands, New Jersey, joined forces to perfect the successful application of airborne radar.

In 1916, what would have been the German's objective in using a diabolical new device to cause innocent swimmers to become lunch for marauding sharks? Perhaps limiting the number of sailors rescued from freshly torpedoed ships. On the mainland, it probably would have been quite inviting for the Germans to plunge the metropolitan region into a state of economic ruin and, at the same time, exert a force that would demonstrate Germany's daunting *dominion* over nature and the United States. Additionally, they could evaluate the American's shark-attack-activated naval power unmasked. Such a horrific state of chaos could create enough of a diversion that the U-boat squadrons could slither right into unprepared northeastern ports to plant explosives and demand immediate surrender. One must also remember that terrorism often does not follow any clearly logical plan, pattern, or meaningful objective. In the 1990s, why did terrorists attempt to flatten New York's World Trade Center or destroy the Lincoln Tunnel?

One thing is certain, bizarre events did take place later in July 1916, which at least lend some evidence to the idea that the unimaginable *could* happen. The Von Rintelen cigar-bomb campaign and the Heinrich Freidrich Albert espionage bungle were not the only malicious German-sponsored acts to fall on U.S. territory during this prewar phase. On July 30, at the Black Tom Island ammunition depot near Jersey City, New Jersey, the largest arsenal of explosives outside Europe, the tranquility of a midsummer night was shattered by a most shocking blast. At 2:08 A.M. on a Sunday, while most residents of the region were fast asleep, the first of three major explosions at Black Tom Island shook the earth. The *New York Times* described the sound as similar to "the discharge from a great cannon." Indeed, the concussion was heard as far away as Philadelphia. Flames from the ignited 200 million pounds of explosives illuminated the sky, drawing curious spectators from miles around. Shock waves knocked out windows in buildings, including skyscrapers throughout the New York area, sending showers of glass cascading toward the trolley-lined streets. Artillery shells kept detonating, and bullets and projectiles whizzed through the air. Shrapnel gouged holes in the Statue of Liberty and into buildings on Ellis Island, then the active entry point to America for wide-eyed immigrants. It seemed as if New York City was being bombarded by a foreign force. Dr. Robert Patterson chuckled during our 1989 talk when I mentioned the Black Tom incident. Dr. Patterson knew a man who was in the vicinity of the sabotage blasts in 1916, and according to Patterson, "he hasn't stopped running yet from all that explosion!"

From the tower of the *Brooklyn Daily Eagle*, in the predawn of the conflagration, was visible a blaze whose flames mounted higher than the Statue of Liberty. The statue itself was clearly outlined against the red sky. Cemetery tombstones toppled, and thousands of dead fish and eels covered the surface of the bay. A ten-week-old boy in Jersey City was fatally tossed from his crib.

German saboteurs had been interested in this site for several months; it was a major supply yard for the Allied countries in Europe. When the smoke cleared from the Black Tom blast, six people had been killed and sixty injured. Vast warehouses, piers, barges, and sixty-nine munition-filled railroad cars were all destroyed, leaving a staggering financial toll of $20 million. In the end, officials determined that German saboteurs, intent on disrupting the arms flow to Britain, Russia, and France, were responsible for the Black Tom disaster. During the same week, an espionage and sabotage ring was uncovered at a German shipping line headquarters in New York. According to Jules Witcover, author of *Sabotage at Black Tom: Imperial Germany's Secret War in America 1914–1917,* "The chaos delivered to New York Harbor in the early hours of July 30, 1916, was the centerpiece of one of the greatest and most cunning deceptions ever perpetrated on the United States by a foreign power."

After the Black Tom blast, President Wilson's competitor for the 1916 election, Charles Evans Hughes, received support from GOP stalwart Theodore Roosevelt. Roosevelt was well aware of Wilson's summer respite at Shadow Lawn and took advantage of the opportunity to present some critical words to the vacationing, neutrality-minded president. Roosevelt took aim at Wilson and declared, "There should be shadows enough at Shadow Lawn; the shadows of men, women, and children who have risen from the ooze of the ocean bottom and from the graves in foreign lands . . . the shadows of deeds that were never done; the shadows of lofty words that were followed by no action; the shadows of the tortured dead."

The Black Tom sabotage, however, was not the first or the last of the major acts of devilish terrorism inflicted on America by Germany in those years. On January 11, 1917, the Kingsland munitions plant in

Lyndhurst, New Jersey, was destroyed by a suspiciously set fire with a $17 million loss. All told, there were about two hundred acts of German sabotage in the United States during this period. In 1939, the Hague Commission awarded some $50 million in damages to the affected American parties of the Black Tom and Kingsland tragedies. Final payments, however, would not be cleared until 1979. The inquisitive shark-attack researcher cannot avoid, at least for a moment, to wonder whether the 1916 attacks should be considered among the two hundred acts of German sabotage/terrorism from 1914–17. Let us not forget that in 1915, Germany's unrestricted U-boat warfare contributed to the torpedo sinking of the *Lusitania* and 128 American deaths. Even though we were not at war in 1916, President Wilson did authorize the government to set up training camps for "vacationing American men" to learn how to become soldiers. Additionally, on April 22, 1915, the Germans warned the United States that any ships crossing the Atlantic which were flying a flag representing a British ally would risk being destroyed. Because of the undeniable acts of terrorism and other worldwide events perpetrated by the Germans during World War I in U.S. territory, it becomes tantalizing to conjure up a scheme to explain the tragic shark events. Although a link between German covert submarine activity and the shark attacks should be considered extremely remote or purely sensational, I will examine an inadvertent connection between the two in a subsequent chapter on modern theories. Among the questions raised is, Could some German scientific development, new to the world in 1916, have accidentally generated a device or mechanism that would account for the numerous dangerous sharks lurking in the vicinity of New Jersey in 1916?

Of all the retraction and change of scientific opinion that the Jersey man-eater influenced, it is likely that the commentary of Robert Murphy represents the most profound modification of views. Let us recall that only three months before the first attack Murphy, as supported by Lucas, was convinced that a shark could not fracture a human bone.

Postcards from the early 20th century depict the Jersey Shore as the northeast's summer playground.

Confusion and debate over the status of German U-boats docked in U.S. ports was evident among the Allies in the summer of 1916. (The Washington Post, *July 10, 1916*)

President Woodrow Wilson. (Courtesy *Entertaining a Nation*)

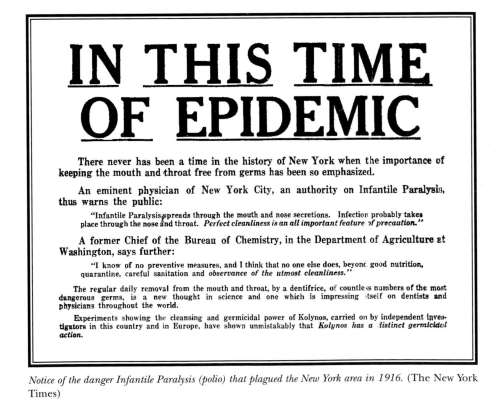

IN THIS TIME
OF EPIDEMIC

There never has been a time in the history of New York when the importance of keeping the mouth and throat free from germs has been so emphasized.

An eminent physician of New York City, an authority on Infantile Paralysis, thus warns the public:

"Infantile Paralysis spreads through the mouth and nose secretions. Infection probably takes place through the nose and throat. *Perfect cleanliness is an all important feature of precaution.*"

A former Chief of the Bureau of Chemistry, in the Department of Agriculture at Washington, says further:

"I know of no preventive measures, and I think that no one else does, beyond good nutrition, quarantine, careful sanitation and *observance of the utmost cleanliness.*"

The regular daily removal from the mouth and throat, by a dentifrice, of countless numbers of the most dangerous germs, is a new thought in science and one which is impressing itself on dentists and physicians throughout the world.

Experiments showing the cleansing and germicidal power of Kolynos, carried on by independent investigators in this country and in Europe, have shown unmistakably that *Kolynos has a distinct germicidal action.*

Notice of the danger Infantile Paralysis (polio) that plagued the New York area in 1916. (The New York Times)

President's Summer Office
To Be at Asbury Park

President Wilson speaking at Shadow Lawn in 1916. (*Courtesy* Entertaining a Nation)

Secretary of the Treasury, William McAdoo.
(Courtesy Underwood & Co.)

The Asbury Park Trust Co. became Wilson's White House during the summer of 1916.

McAdoo's Spring Lake residence on the corner of Passaic and First Avenue.

The Engleside Hotel, Beach Haven, NJ, 1916. (Courtesy John Bailey Lloyd)

Charles Vansant would have arrived at Beach Haven by train from Philadelphia. (Courtesy John Bailey Lloyd)

Charles Vansant (bottom right) with his classmates from the University of Pennsylvania. (Courtesy University of Pennsylvania)

The lobby of the Engleside Hotel. Vansant expired on the manager's desk. (Courtesy John Bailey Lloyd)

Alexander Ott with son, Jackie, pictured here in the 1994 Sports
Illustrated *swimsuit issue. (Courtesy Jackie Ott)*

DEATH STRUGGLE WITH SHARK
DESCRIBED BY ONLOOKERS

Persons on Strand at Beach
Haven, N. J., Vainly Tried
to Save Man's Life

FURTHER details concerning the tragic
death of Charles Epting Vansant became known yesterday, when his body was
brought from Beach Haven, N. J., to the
home of his father, Dr. E. L. Vansant, at
4038 Spruce street.

Vansant died on Saturday afternoon at
the Engleside Hospital, at that seashore
resort, after a terrific battle with a nine-
foot shark in the surf. He had gone from
here with his father and two sisters to
spend the week-end there, and on Saturday
afternoon was in the surf only a few yards
from shore when attacked by the man
eater. He was playing with a dog at the
time, and onlookers who heard his cries
thought them only part of the game.

Vansant tried to get to shore with the
jaws of the shark clutching his leg. In the
shallow water persons on shore saw the
man eater's fin and ran to help him. Led
by Alexander Ott, a champion swimmer
and member of the American Olympic
team, they drove the shark off and carried
its victim to the beach. His leg had been
torn from the thigh to the knee. Physicians were called at once, but he died an
hour and a half later.

Vansant is well known here, having been
graduated from the Episcopal Academy in
1910 and from the University of Pennsylvania in 1914. At the latter institution he
was a member of the Glee Club, assistant
business manager of the Record and was
on the business staffs of the Punch Bowl
and the Red and Blue. At the time of his
death he was connected with Folwell Brothers & Co., of this city. He was the only
son of Dr. C. L. Vansant, whose offices are

at 1929 Chestnut street. His father is a
member of the Union League and is a
prominent nose and throat specialist.

The funeral will be held from the home
here at 11 o'clock on Wednesday morning.
The Rev. Dr. Archibald McCallum, of the
Walnut Presbyterian Church, will officiate.
Interment will be in South Laurel Hill
Cemetery.

CHARLES EPTING VANSANT

One of many notices in the local papers of Charles Vansant's death
(Philadelphia Public Ledger).

Charles Vansant's death certificate was the most legible of the attack victims. Cause of death reads, "Hemorrahage from Femoral Artery, left side" and "bitten by a shark while bathing."

Vansant's headstone in South Laurel Hill Cemetery.

BATHERS NEED HAVE NO FEAR OF SHARKS

Fish Expert Declares One That Killed Swimmer May Have Sought to Attack Dog.

Many local papers, including The New York Times *(below), sought to downplay the idea that Vansant was attacked by a shark.*

DIES AFTER ATTACK BY FISH

C. E. Vansant Had Been Bitten While Swimming at Beach Haven.

The Essex & Sussex Hotel, Spring Lake, New Jersey, 1916.

SKIPPERS SAY SEA IS ALIVE WITH SHARKS

Incoming Mariners Report Great Schools in Atlantic Steamship Lanes.

Sea captains were reporting seeing many more sharks than was usual off the Atlantic coast that summer.
(Philadelphia Eagle)

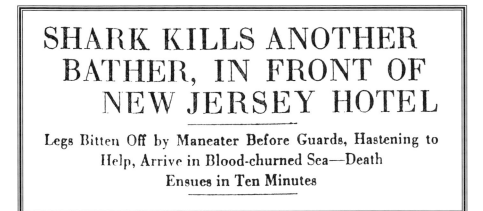

SHARK KILLS ANOTHER BATHER, IN FRONT OF NEW JERSEY HOTEL

Legs Bitten Off by Maneater Before Guards, Hastening to Help, Arrive in Blood-churned Sea—Death Ensues in Ten Minutes

Headlines announced the shocking news that a second bather was attacked. This time there was no mistake that a shark was the culprit. (Philadelphia Inquirer)

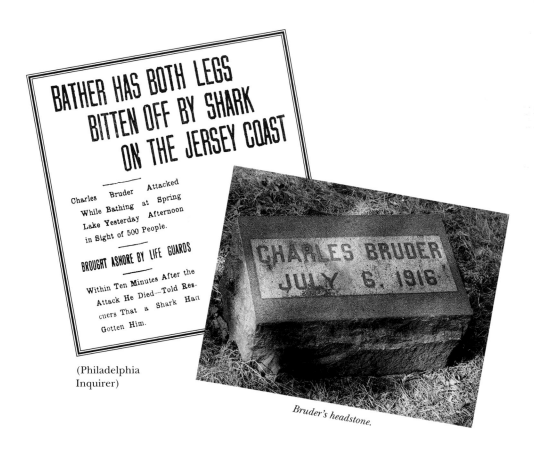

BATHER HAS BOTH LEGS BITTEN OFF BY SHARK ON THE JERSEY COAST

Charles Bruder Attacked While Bathing at Spring Lake Yesterday Afternoon in Sight of 500 People.

BROUGHT ASHORE BY LIFE GUARDS

Within Ten Minutes After the Attack He Died—Told Rescuers That a Shark Had Gotten Him.

(Philadelphia Inquirer)

CHARLES BRUDER
JULY 6, 1916

Bruder's headstone.

William Schauffler, M.D. (Courtesy New
Jersey Medical Society)

William Trout, M.D.

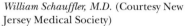

Letter from William W. Trout, M.D. to District Coast Guard reporting the extent of Bruder's injuries. He and Dr. Schauffler examined Bruder on the beach just after the attack. (Courtesy
National Archives)

Postcard from a honeymooner in Asbury Park, dated July 8, 1916. It reads "... They have screened it [the bathing area] in and it is patrolled by boats since the scare of a shark biting off the legs of a man a few beaches above here the other morning. The man died. Since then a great many bathers are rather scarce ..."

Postcard from a female bather in Asbury Park, dated August 17, 1916 jokes, "I'm safe from the sharks; they are the <u>man</u>-eating variety."

Ads announcing shark-proof bathing nets aggressively sought to allay fears and keep bathers coming down to the shore.

PHILADELPHIA, SATURDAY MORNING, JULY 8, 1916.

MOTORBOATS SEEK SHARKS OFF NEW JERSEY COAST BEACHES TO PROTECT SURF BATHERS

NINE FOOT SHARK LANDED AT STRATHMERE N.J.

Along with erecting netting, beach officials began patrolling the bathing areas in boats armed with rifles and shotguns. (Philadelphia Public Ledger)

Postcard from the early 20th century depicting Matawan's main street.

Presenting
The Cecil

The ultra nifty style shown here tailored for <u>you</u> to your individual measurements and to your <u>absolute</u> satisfaction from any of the hundred splendid fabrics in my store, by The Royal Tailors, Chicago, New York, at prices from $16 to $38.

The Royal Rigid Guarantee which is lived up to in every detail:

A view of Matawan creek in 1900. The attacks on Stillwell and Fisher occurred near this location sixteen years later. (Courtesy of Maurice Cuocci Collection)

The Royal Tailors are <u>proving</u> their worth in this town, as everywhere else, and in fairness to yourself it will be worth your while to visit this store before you close any deal for your suit.

We are doing a <u>splendid</u> business, and the best of it is that old customers are coming in for new orders—a splendid recommendation.

This Garment is Guaranteed to Fit You Perfectly

¶ If you are not pleased with it in every respect we ask you not to accept it, not to pay one penny

W. STANLEY FISHER

Authorized
Resident
Dealer

Genuine
Block
Matawan

Advertisement for Stanley Fisher's shop in Matawan posted only two months before Fisher's death. (Matawan Journal)

Lester Stillwell. (New Brunswick Times)

Stillwell's headstone.

Stillwell's death certificate. Cause of death reads "Bitten by a shark while bathing in Matawan Creek" and "Hemorrhage and Shock." Lester's father listed "school" as Lester's occupation.

Redbank, N. J., July 17, 1916.

Superintendent of Coast Guard,
    Asbury Park, N. J.

Dear Sir:

    The following is the history of W. Stanley Fisher as
requested:

    W. Stanley Fisher, age 24; residence, Matawan; occupation,
tailor. Admitted to Monmouth Memorial Hospital, Long Branch,
N. J., July 12, 1916 at 5:30 P. M. Was injured at the old pier
in Matawan creek, Matawan, N. J., while endeavoring to recover
the body of Lester Stilwell, who had been killed by a shark.

    Fisher dived into the water and had found the body, when
he was seized by the right thigh. The muscles were torn out.
Fisher was taken out of the water by friends nearby in a boat,
but bled profusely before medical assistance could be summoned.
A tourniquet was applied and he was hurried to the hospital.

    Condition on admission: suffering from shock and pulseless
from loss of blood. Did not rally to stimulants and saline
transfusions. The outer side of the right thigh was denuded
from three inches below the great trochanter to two inches
above the knee, all of the muscles and tissues being completely
removed, and only a third of the muscular tissue on the inside
of the thigh remaining. Bone not injured.

    Died at 6:35, one hour after admission.

Yours respectfully,

Examined and forwarded, JUL 18 1916 13

John L. Cole
    District Superintendent

Edwin Field.
Visiting Surgeon to
Monmouth Memorial Hospital

Stanley Fisher

*The report to District Superintendent Cole from Edwin Field, M.D.,
the doctor on duty at Monmouth Memorial Hospital in Long Branch
when Fisher arrived for treatment.* (Courtesy National Archives)

Edwin Field, M.D. (Courtesy
New Jersey Medical Society)

*Almost the entire town of Matawan came down to the train station to see Fisher being loaded on to the
train and rushed to Long Branch.* (Courtesy Dorn's Historic Photos)

The Fisher family plot with Stanley's headstone in the background.

Matawan United Methodist Church and a plan for the Bethlehem Window erected in memory of Stanley Fisher.
(Courtesy Walter Jones)

Little Johnny Smith, Fisher's messenger boy. This photo was taken around the time of the attacks.

Albert O'Hara. (The Brooklyn Reporter)

Joseph Dunn. (New Brunswick Times)

MOTHER OF SHARK VICTIM PAID $7,500

Ralph Gorsline, local agent for the London and Lancashire Indemnity Company, has paid over to the mother of W. Stanley Fisher, who gave his own life in a vain attempt to rescue Lester Stillwell, from the jaws of a shark in the Matawan Creek two weeks ago, the sum of $7,500. The victim's mother was the beneficiary under the insurance contract made with the company. Mr. Gorsline, who was closely acquainted with Mr. Fisher, induced his family to take out the accident policy only last February.

The principal of the insurance was paid-over-to the beneficiary of the company before the heroic young man was buried. When Stillwell down in the creek by the man-eating shark, Fisher promptly jumped in and attempted to rescue him. He was badly bitten and died a few hours afterward in the Long Branch hospital.

Article in a local paper announcing the insurance policy paid to Fisher's mother. (Asbury Park Press)

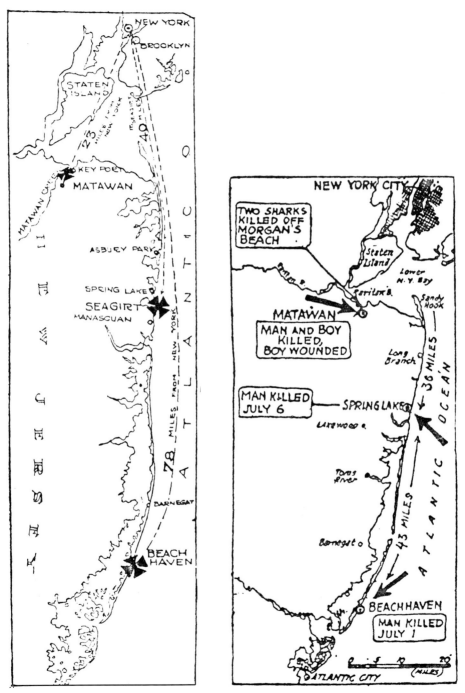

Local papers published maps showing the locations of the five attacks.

FINAL
SPORTS

THE EVENING MAIL

FINAL
SPORTS

80TH YEAR. NO. 165.   NEW YORK, THURSDAY, JULY 13, 1916.   WEATHER—Partly cloudy tonight and ONE CENT.
                                                                    Friday. Not quite so warm. See page 2

HUNDREDS SEEK TO SLAY SHARK;
THINK MONSTER TRAPPED IN CREEK

Matawan citizens took justice into their own hands. They hunted the killer sharks by any means, including shotgun and dynamite. The men in the rowboat (above) are probing for Lester Stillwell's corpse. (Both photos courtesy Brown Brothers)

Dozens of political cartoons that summer used the shark as a symbol of evil.

SHARK FOOD

Nelson Harding

GETTING HIS FOOT IN IT—By Brinkerhoff.

To the Mayor —
Matawan N.J.

Dear Sir: —

After reading the account of the terrible sharks tragedy it occurred to me that I could be of assistance to you & do a great deal of good to the community by killing out the monsters for you as I have made a life time study & business of capturing & killing sharks & other big fish —

I know I can rid your coast of the dangerous sharks if your city will cooperate with me —

I can furnish you with all the reference you wish as to my ability in that line.

Hoping to hear from you soon in regard to the matter.

I remain,
Very truly Yours
Capt. Chas. H. Thompson
Box 378

Re-Route Miami Fla —
July 24th '16

July 13 1916

Shark Terrors can be promptly eliminated wire for representative our expense

Cyclone Fence Company
256 Canal St.
New York City

Dated New York NY 1250 PM 13
To Mayor of Matawan NY
Matawan NJ

Letters and telegrams poured in to Matawan suggesting various ways to rid the town of its shark menace. (All pieces courtesy Clark C. Wolverton)

Dear Sirs I hope the following invention might be of some assistance in catching the sharks.

spring

to Batteries or lighting supply 110 V.

Box or float

Ball of Meat Tied with strong piano wire

2 wires from exploder.

Box, or can of Dynamite with electric exploder

When the Shark pulls the bait he pulls up the spring & makes contact with the electric circuit which sets off the dynamite I hope you may use this
Alex Murdoch Jr
47 Lincoln Ave
Lansdowne Pa.

PHILADELPHIA PA
JUL 14
8 AM
1916

PENN SQUARE
-STATION

Village Authorities
Matawan
New Jersey —

Captain John Cottrell (seated) and his nephew, John (standing, right). (Courtesy Malcolm Barhenburg)

SHARK KILLED
AT KEYPORT;
HUMAN BODY INSIDE
TERROR SLAIN NEAR
SCENE OF TRAGEDY

Erroneous report that human remains were found inside a large shark caught near the scene of the Matawan attacks (The Newark Star Eagle).

BODY OF BOY VICTIM OF SEA WOLF FOUND AT MATAWAN

Torn and Mangled Corpse of Lester Stillwell Rises to Surface Near Where Shark Dragged Youth to Death---Mother Is Overcome by Shock.

BELIEVE THAT FOUR MONSTERS ARE TRAPPED

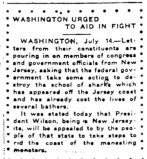

WASHINGTON URGED TO AID IN FIGHT

WASHINGTON, July 14.—Letters from their constituents are pouring in on members of congress and government officials from New Jersey, asking that the federal government take some action to destroy the school of sharks which has appeared off the Jersey coast and has already cost the lives of several bathers.

It was stated today that President Wilson, being a New Jerseyite, will be appealed to by the people of that state to take steps to rid the coast of the maneating monsters.

Lester's body is recovered in the creek on July 14. (Asbury Park Press)

BODY FOUND; SHARK HUNT IS RENEWED

Mutilated Corpse of Lester Stillwell Discovered Floating Near Scene of Tragedy.

After Stillwell's body is recovered, previous reports about his corpse being found elsewhere were discounted. (Newark Star-Eagle)

U. S. WAR ON SHARKS

Wilson and Cabinet Make Plans to Prevent More Tragedies.

COAST GUARDS TURN HUNTERS

Federal Cutters Also Are Ordered to Fish for the Monsters.

Bureau of Fisheries Issues Warning, but Admits Inability to Prevent Attacks—Bacharach Asks Congress for $5,000 to Aid the Campaign. Theories of Scientists for the Presence of Maneaters on the Coast.

Front page headline from The Washington Post, *July 15.*

SIX-FOOT SHARK CAPTURED WITH ROD AND LINE IN SURF AT BEACH HAVEN, AND ONE OF ANGLERS WHO LANDED IT

E.F. Warner of Field and Stream *magazine landed a six-foot shark with rod and reel at Beach Haven, New Jersey, a few days after the attacks.* (New York Herald)

FEDERAL AGENT HERE TO PLAN SHARK FIGHT

Capt. G. L. Carden ot Revenue Cutter Mohawk, After Local Conference Decides ' Wire Netting Is Best Form of Protection.

Captain Commandant G.L. Carden inspired the July 15th conference at Asbury Park, led by McAdoo, to try and protect bathers on the Jersey Shore. (Asbury Park Press)

Letters from coast guard station keepers in response to the July 15th conference affirming that nets had been installed at New Jersey beaches. (Courtesy National Archives)

NET AT ASBURY PARK PROTECTING BATHERS

Front page photograph announcing that nets had been installed at Asbury Park. (Newark Star-Eagle)

250-POUND SHARK CAPTURED; RECOVER BOY VICTIM'S BODY; U. S. COAST GUARDS TO AID

Belford Fishermen Kill Man-eating Sea Beast Found In Nets---Stillwell's Body Found Near Scene of Tragedy.

SHARK HUNT CONTINUES THO WITH FAINT HOPES

The capture of a 250-pound man-eating shark in the nets off Belford, discovery of the torn and mutilated body of Lester Stillwell, boy victim of the sea monster, and news from Washington that Secretary of the Treasury McAdoo had directed the Coast Guard to aid in the fight on the dangerous fish, were the outstanding developments yesterday in the effort that is being made to prevent any repetition of the terrible tragedies that took place in Matawan creek two days ago.

Front page headline announcing the capture of the Belford shark. (Asbury Park Morning Press, *July 15*)

Harlem Man In Tiny Boat Kills
A 7½ Foot Man-Eating Shark

Beats It to Death With Broken Oar. Directly Off Matawan Creek,
Where Two Bathers Were Attacked and Killed By Sea-Tiger
Last Week—Examination By Director of Museum of Nat-
ural History. Shows Human Bones in Shark's Stomach.

Schleisser shark is captured in Raritan Bay. (Bronx Home News)

Schleisser pictured with the famous Great White. (Bronx Home News)

Man-Eating Shark, Displayed in Home
News Window, Thrills Many Thousands

Headline announcing the display of the Schleisser shark at the Bronx Home News *office.*

DEPARTMENT OF COMMERCE   M
BUREAU OF FISHERIES
WASHINGTON

ADDRESS ALL COMMUNICATIONS TO
COMMISSIONER OF FISHERIES
WASHINGTON, D. C.

RECEIVED   July 14, 1916.

JUL 15 1916

DIRECTORS DESK.

Dr. F. A. Lucas,

    American Museum of Natural History,

    New York City.

Dear Doctor Lucas:

    I have observed from the newspaper press that Mr. Nicholls has been sent to Matawan to make inquiries concerning the shark or sharks which have recently attacked persons at that place. We have had a number of inquiries as to whether it was our purpose to make an investigation of this kind and have replied to the effect that it was not, as Mr. Nicholls, who is already on the ground would be able to gather all the facts that it will be possible for us to get.

    Will you kindly arrange that Mr. Nicholls should send us as promptly as possible anything definite that he may learn. I assume, of course, that there is no question but that the recent fatalities were due to sharks.

    With best regards,

                Very sincerely,

                    H.F. Moore

                    Acting Commissioner.

Dear Mr. Moore:

    I hasten to reply to your letter of July fourteenth, in order that you may know that we shall be very glad indeed to send you any information that Mr. Nichols may obtain.

    There seems to be no question that the fatalities which have occurred are due to sharks, but the species is still undetermined.

    With kind regards,

                Faithfully yours,

                        Director.

Mr. H. F. Moore,

Acting Commissioner, Bureau of Fisheries,

Washington, D. C.

DEPARTMENT OF COMMERCE   M
BUREAU OF FISHERIES
WASHINGTON

 ALL COMMUNICATIONS TO
SIONER OF FISHERIES
WASHINGTON, D. C.

    July 18, 1916.

RECEIVED

JUL 19 1916

DIRECTORS DESK.

Dr. F. A. Lucas,

    American Museum of Natural History,

    New York, N. Y.

Dear Doctor Lucas:

    I thank you for your letter of July 17, and for your willingness to furnish this office with any reports which may be made by Mr. Nichols in regard to the shark plague on the New Jersey Coast.

                Very truly yours,

                    H.F. Moore

                    Acting Commissioner.

              August the seventh
              Nineteen hundred sixteen

Dear Mr. Moore:

    I am sorry to say that there is nothing special to report in regard to the shark situation on the New Jersey coast, as Mr. Nichols was unable to get any information other than that published in the papers, and there have been no instances of any attacks since that at Matawan creek.

    On July 14, off South Amboy, a 7 1/2-foot specimen of the white shark was captured. This Mr. Nichols tells me is the first instance of this species having been taken within fifty miles of New York, although it had previously been caught at Provincetown. This particular specimen had in its stomach parts of the left forearm and left upper rib of a robust man. While the "physicians" reported that the bones were those of a boy, it was incorrect. From the size of the shark and the condition of the bones, it is probable that the bones were taken from a body that had been dead some time and not the result of any active attack. The reported injuries to Bruder, who was the first to be killed by a shark, would indicate that that particular shark was not a very large one. The detailed description is as follows:

    "Charles Bruder was seen by me, (Dr. W. G. Schauffler) about 1/2 hour after he was brought to shore. His right leg was gone from half way between the knee and ankle. The tibia and fibula being torn and jagged, the muscles torn and hanging in shreds from just below the knee. His left foot was missing, the lower end of tibia and fibula denuded of flesh and the flesh and muscles torn and jagged from the middle of the leg. There was a circular gash about the left knee extending down to the femur. On the right side of the abdomen, a piece of skin, muscle and fat was gouged out about as large as an apple. The peritoneum was not opened."

Letters between Dr. Frederic Lucas (above, right) and Commissioner of Fisheries, H.F. Moore. (Courtesy Department Library Services, the American Museum of Natural History)

Dr. John Nichols. (Courtesy Department Library Services, American Museum of Natural History)

Dr. Robert Murphy. (Courtesy Department Library Service, American Museum of Natural History)

The American Museum of Natural History in New York, as it would have looked in 1916. (Courtesy Department Library Services, American Museum of Natural History)

Before the end of July 1916, Murphy was conveying facts about the white shark that hold true even by today's research standards. By this time, Murphy and his colleagues were most likely putting more credibility in the shark stories they had heard and read about from undeveloped lands. In the July *Scientific American,* he reported: "The white shark is perhaps the rarest of all noteworthy sharks . . . their habits are little known, but they are said to feed to some extent on big sea turtles. . . . Judging from its physical make-up, it would not hesitate to attack a man in the water." In reference to the ability of a shark to chew through bone, Murphy felt that it was now such an obvious capability, it rendered the question itself unimportant, saying, "because it is evident that even a relatively small white shark, weighing two or three hundred pounds, might readily snap the largest human bones by a jerk of its body, after it has bitten through the flesh."

Murphy's bold acknowledgment of the true power in a shark's jaws tells much about the distribution and acceptance of research data during that period. It was only after the extraordinary attacks in New Jersey that Lucas, Murphy, Nichols, and many other scientists admitted that a shark could perpetrate such acts against man. After some digging, Jim Foley, a coenthusiast of Jersey Shore history, presented me with an interesting excerpt from the 1860 *Annals of Philadelphia and Pennsylvania in the Olden Time.* The excerpt clearly shows that the naturalists of 1916 should have been aware of sharks' jaw prowess. The passage represented a diary entry from August 8, 1822, of Charles Woolston, who made a business of using a tremendous seine net in the Delaware Bay to catch porpoises and sharks. The passage reports a catch of six sharks in a single one-day haul.

> One or two of the sharks were eight feet long. They had great muscular vigor after their heads were off; and their heads—for several hours after they were off—if stood up on their base, and a stick thrust into their throat, would snap violently and with much strength their teeth into the wood. It is strange that

these sharks occupy the same waters in which we bathe, and
yet never molested us. Even when some go out swimming and
floating beyond their depth.

Perhaps the age of the source explained why scientists treated such information with skepticism. Possibly its remote location (in a local history book), even with the Dewey decimal system (established in 1876),
meant that the scientists simply could not find the data.

Not only did the capture of the Raritan Bay shark inspire Murphy to
change his general and specific views on attacks, but that particular
white shark seemed to convince Murphy and Nichols of much more
about the 1916 events themselves. In a *Newark Star-Eagle* article that appeared in late July 1916, the author of the article pursued some reso-
lution to the shark crisis and quoted Murphy, who stated:

> . . . the occurrence of the white shark near New York, being al-
> most unprecedented as the attacks on bathers which hap-
> pened simultaneously, the capture of a specimen by Mr.
> Schleisser confirms our [his and Nichols's] belief that the
> white shark was responsible for the casualties. Indeed, that a
> single fish was at the bottom of the successive attacks at Beach
> Haven, Spring Lake, and Matawan.

It is quite interesting, and uncommon, that the writer of the piece
who documented Murphy's statements also criticized other press writers for inaccurate reporting. The writer specifically condemns the
press for attributing outrageous theories to shark authorities. One
theory proposed that ordinary ground (sandbar) sharks, rendered
desperate through hunger, were the cause of the recent loss of life in
New Jersey.

Regardless of the debate over the cause or the true culprit in the
1916 attacks, no further summer shark-related events of significant
human consequence occurred after the Schleisser catch on July 14. In

a letter to H. F. Moore on August 7, Dr. Lucas refers to the injuries detailed by Colonel Schauffler for the Bruder attack but also comments on the human items recovered from the Schleisser white shark. Despite Lucas's confirmation that the shark's stomach contents were human, he clearly demonstrates his personal reluctance to admit that the Raritan Bay white was responsible for any active attack in 1916. Astonishingly, Lucas even refuses to agree that Charles Vansant was killed by a shark. His letter stated, in part:

> On July 14, off South Amboy, a 7 1/2-foot specimen of the white shark was captured. This, Dr. Nichols tells me, is the first instance of their species having been taken within fifty miles of New York, although it had been previously caught at Provincetown [Massachusetts]. This particular specimen had in its stomach, parts of the left forearm and left upper rib of a robust man. While the 'physicians' [quotation marks unexplained] reported that the bones were those of a boy, it was incorrect. From the size of the shark and the condition of the bones, it is probable that the bones were taken from a body that had been dead some time and not the result of any active attack. The reported injuries to Bruder, who was the *first* to be killed by a shark, would indicate that that particular shark was not a very large one.

An appreciative Deputy Commissioner Moore sent back an immediate return correspondence and appropriately responded, "The excitement in this matter appears to have died down, much to the relief of this office, and I hope nothing will occur to resuscitate it."

In Asbury Park, during early August 1916, another tremendous meeting was assembled at the convention hall to discuss one topic: sharks. While President Wilson, his Cabinet, and the New Jersey State Chamber of Commerce tried to downplay the gruesome July events and promote enjoyment of the shore's true splendor, the marine scientists were busy regrouping and came in hopes of achieving some type of smoke-clearing consensus. To the professional analyst and the layperson, the general sense of things was that New Jersey and the United States experienced one of those freakish natural events. It was the equivalent of an unexplained famine, plague, or seldom-seen volcanic eruption. The worsening world war would later be referred to as the "war to end all wars," and in many ways, this was the shark attack to end all shark attacks. The *Titanic* of shark attacks.

During that bizarre summer, beach vendors were not the only ones scratching their heads. The American Aquatic Union Swimming Committee was at a loss as to where the outdoor distance swim competition would be held. As they mentioned, "Few athletes would chance the loss of life to retain possession of an outdoor swimming title. . . . These swimmers are a courageous lot . . . but they know just the amount of danger they are compelled to face every time they plunge into the deep. They know that to encounter a shark in deep water is certain death." The answer was to get the competitors to the large available pools. I wonder if the late Hermann Oelrichs would have curtailed his long-distance swims?

In Matawan, things were returning to normal. Visitors began flocking again to the shore, but now they were content to utilize the large public pools. Johnson Cartan told me that he enjoyed the pools at Long Branch and Asbury Park, and he would often take the train ride there with Lester Stillwell's brother. Cartan would also say, "When we arrived there, we didn't think about going in the ocean. The ocean did have nets, though."

Despite some return business to the seashore in late summer, New Jersey lost more than a million dollars in revenue and cancellations, which is equivalent to approximately sixteen million dollars today.

On the subject of pools and scientific pursuit of the shark problem, a letter ended up on the desk of the U.S. surgeon general, then made its way to the Coast Guard service, then through the hands of the president's personal secretary, and finally made its way to the eyes of President Wilson. The letter was typewritten (unusual for the time). Though its message was initially a bit shaky, it rallied to make a point that prompted its circulation to the officials. The letter, from Frank Bryan of Rockaway, New York, started out by saying, "I am not a crank or a fanatic, but I appeal to you because I fully believe that you alone will give my idea the consideration which I think it deserves." Bryan felt that the dangerous sharks were "here to stay" and that the futility of trying to kill them was especially disastrous when it was combined with the infantile paralysis epidemic that was raging. Bryan pointed out that ocean water is one of the best antiseptics and preventatives of all. The letter made the statistical argument that the number of cases of the deadly paralysis were remarkably small near the sea. He related the healing qualities of the sea with its iodine and halogenated contents. Bryan pushed for the immediate construction of a uniform steel-railed apparatus to allow mass bathing along the shore. He dramatically mentioned that the same manufacturers of steel shells for the "destruction of life in Europe might then be used for the preservation of life in America."

There is no doubt that Bryan's argument had some merit. Every modern physician is aware of the soothing, rehabilitative potential of swimming and bathing. Additionally, the seawater's salinity and natural iodine content (from marine algae) make it a powerful anti-infective agent. Today, medical personnel commonly use saline solution to cleanse wounds. The sea does, however, possess its own brand of nasty gram negative bacteria, which can, for example, contribute to sinus infections, ear infections, and wound infections, especially in cases of injury from sea life (e.g., urchins, clam shells, stingrays, etc.). The fact that Bryan was touting the sea's antipolio correlation may not have been attributable to the ocean's antiviral capacity but rather to the fact that bathers living near the sea spend less time in enclosed communal swimming locations such as pools or ponds.

Of all the characters from 1916, I would say I am most disappointed not to have had a chance to talk with Joseph Dunn. Even with the assistance of Jerry Hourihan's grandson, I was unable to locate Joseph's Cliffwood relations. I searched the census records of New York in 1916 and even made several phone calls to the multiple Dunns of Staten Island. (I had gotten a tip that the Dunns moved out of the Manhattan/Brooklyn area years before and headed to nearby Staten Island.) I even visited Joseph's 1916 boyhood home.

Fortunately, what I do know about Joseph and his recovery is pleasant. On September 15, 1916, more than two months after the attack, Joseph was finally released from St. Peter's Hospital into the care of his family. On that day, his brother, Michael, held two sacks of get-well cards sent by well-wishers from throughout the country. Hospital staff and reporters were all smiles as they watched the only survivor of the summer terror make his way back to the real world. Joseph stated that he would write a response to each of his well-wishers, starting with President Wilson. It was already September and school was in session, but Joseph would wait out another semester before returning as a hero to St. John's Academy. Joseph left the hospital with two legs, a limp that he would outgrow, and a pair of dark wooden crutches. He was in need of much more rehabilitation, but eventually he would walk soundly without an assistive device. To provide the best environment for his full recovery, Joseph was to take the train to Matawan, then continue on to his aunt's bayshore home in Cliffwood. Understandably, Joseph's mom did not want the boy to return to New York until the infantile paralysis epidemic was over.

When Joseph disembarked the train at Matawan that late afternoon, en route to his aunt's house, he encountered a small group of well-wishers, including Mary Anderson and the pensive-looking Albert O'Hara, one of Lester Stillwell's closest buddies. Albert was still awash in survivor's guilt, but the two boys did exchange brief words. Albert was apparently surprised to see the upbeat demeanor displayed by the limping Joseph. Young Joseph did have lifelong scars on his lower leg but he would remain an inspiration to a host of people. When asked how he finally recovered from his shark-filled nightmares, Joseph mentioned that he felt it would be selfish to feel sorry about what had happened to him. He said, "Four other people got killed by that shark, and I think the least I could do is feel fortunate I am able to live my life." I can't help but think that Ally felt he should do the same.

The jaws of Carcharadon Megalodon, the 45-foot ancestor of the great white.

CHAPTER 7

The New Generation of Shark Hunters

A second generation of theories, which postulated that the lone Raritan Bay white caused the entire 1916 catastrophe, were likely precipitated by the writings (the original theory) and comments of Robert Murphy, John Nichols, and a handful of other authorities of the original generation. In the decades following, any semidetailed rendering of the 1916 attacks found in popular literature would also invariably suggest that the Raritan Bay great white was the culprit. This is not to say that there was not some amount of disagreement with respect to the details. The 1961 book *Shark!*, by Thomas Helm, included a description of the 1916 events, but the author remained noncommittal about the reason(s) for the unusual occurrence. Helm did report that one of the bones recovered from the Schleisser shark was positively identified as the shinbone of Charles Bruder, but he probably meant to say "a boy"—Lester Stillwell. No one ever said that the bones found in the Raritan Bay white had any specific connection to Charles Bruder, but with this assertion, along with the fact that the attacks stopped after the Schleisser catch, Helm seemed to suggest that the white shark caught was undoubtedly the shark responsible for the string of attacks. In his final statement on the attacks, Helm wrote: "The reason why these five people were attacked will never be known for sure, but the records remain for the serious students to study and ponder."

Two years after Helm's book, a most impressive work on sharks, skates, and rays was released called *Shadows in the Sea*. The authors—Harold McCormick, Thomas Allen, and William Young—devoted a complete chapter to a very detailed account of the 1916 attacks and offered some thought-provoking speculation. They, too, seemed to advocate the great white theory, at least in reference to the species responsible, but not necessarily that one shark was the culprit.

In the 1964 publication *Danger—Shark!*, Jean Campbell Butler implied that the New Jersey attacks constituted an open-and-shut rogue-shark case. He said, "The classic example of rogue-shark behavior in the United States was of this long-range cruising kind. . . . The route of attacks consistently moved in a northerly direction. It terminated when a shark was caught at the northern end. The shark, positively identified as a great white shark, still had in its stomach a mass of human flesh and bones." Similarly, in *The Nightmare World of Sharks* Joseph Cook and William Wisner wrote: "In one of the worst shark villainies ever to occur in the United States, the assassin was proved to be a white shark. . . . A series of five attacks . . . and all were blamed on the same cruising man-eater. . . . After this [Raritan Bay white] capture the attacks promptly ceased, and there was little doubt but that the white shark had been the terror."

Eleven years after the spectacular 1963 account in *Shadows in the Sea*, and after innumerable shark-catching exploits by Frank Mundus, of Montauk, New York, Peter Benchley came out with the novel *Jaws* in 1974. Richard Ellis believes that of all the source subjects from which Benchley drew for *Jaws*, Mundus and the 1916 attacks are at the top of the list. In a letter of thanks to me from Benchley, in response to a complimentary copy of *In Search of the "Jersey Man-Eater,"* Benchley referred to the 1916 attacks as "epochal."

In *Sharks: Attacks on Man,* Dr. George A. Llano departed from the usual rhetoric regarding the accusation of the Raritan white. Instead, he ventured:

> One of the most surprising aspects of the Matawan Creek attacks was the distance from the open sea. Elsewhere in the book are accounts of well-documented shark attacks at Ahwaz, Iran, which is 90 miles up river from sea. It may also be of interest to note that sharks live in Lake Nicaragua, a freshwater body, and in 1944 there was a bounty offered for dead freshwater sharks, as they had "killed and severely injured lake bathers recently." The sharks grow to more than four feet in length.

Here Llano is obviously alluding to the possibility or probability that a bull shark (the U.S. East Coast regional variation of the Lake Nicaragua shark) should be considered in the 1916 Matawan, New Jersey, attacks.

As I've mentioned earlier, Richard Ellis had the opportunity to examine the voluminous 1916 documents gathered by the senior author of *Shadows in the Sea,* and he used those sources to formulate his own opinion of the confounding attacks. In *The Book of Sharks,* Ellis questioned the likelihood of accusing one shark—covering the range of forty-five miles in five days (Beach Haven to Spring Lake) and another thirty miles in six days (Spring Lake to Matawan)—of all five attacks. Ellis also pointed out that great whites rarely, if ever, travel into fresh inland waterways. He found it illogical to determine that the white caught in the Raritan Bay could have been the Matawan Creek culprit, since the creek was closed off with chicken-wire mesh netting shortly after the attacks.

In respect to the identity of the Matawan Creek attacker, Ellis found the likely culprit to be a bull shark (*Carcharhinus leucas*). He drew this conclusion in view of the bull's tendency to enter freshwater (the regional variety, *Carcharhinus nicaraguensis,* may even be totally adapted to a lifetime of freshwater existence), its nasty reputation for attacking people (especially off the Natal coast of South Africa), and because its common attack display of repeated biting and slashing (common to all sharks of the Carcharhinidae) closely matches the attack pattern at the Matawan Creek. Ellis also had speculated that the seven-foot, 230-pound shark caught in the creek (or at the mouth of the creek) by Captain Cottrell six days after the attacks could very well have been a bull shark. The fact that Cottrell's shark was the only shark caught very close to the creek was significant to Ellis because it supported the contention that only a bull species was or could be found in that freshwater. Ellis finished his original argument for the bull by noting that though the bull shark is not common in New Jersey, it does appear more frequently than the white.

Ellis's 1975 bull shark theory was supported by strong reasoning, and he skillfully utilized the knowledge and documentation he had before him. Others have either directly or indirectly shown support for Ellis's assertions regarding the bull shark. In Rodney Steel's *Sharks of the World,* Steel pointed to the traditional white shark scenario as the ending to the 1916 attack saga, but later in his book, without specifically referring to the 1916 case, he provided a rather elaborate explanation of the bull shark's unique capacity to physiologically adjust to freshwater. Steel's discussion of this capability and physiology pointed to a fairly sophisticated adaptation system in the bull. He wrote: "The facility with which *Carcharhinus leucas* transits from the salt waters of the Caribbean to the fresh water of Lake Nicaragua has been a source of considerable interest, because sharks are geared to life in a marine environment." Steel explained that the shark's excretory system normally retains nitrogenous waste, largely in the form of urea, in the bloodstream, thus maintaining the animal's total osmotic pressure at a level comparable to that of the surrounding salt water and preventing dehydration. "The lake Nicaragua sharks," he continued, "it seems, retain less urea in their blood once they leave the sea and also reduce the level of salts in their body fluids." The end result is the ability to maintain a viable osmotic balance between their bodies and the freshwater of the river system and the lake, despite incurring a slightly higher total water content and a smaller ratio of extracellular to intracellular fluid. Steel's conclusions seemed to have at least some foundation in the work of T. H. Lineaweaver and R. H. Backus. In the chapter "A Matter of Controversy," found in their *The Natural History of Sharks,* they provided an early and commanding grasp of the inexplicable and unrecognized qualities of the bull shark.

There is further data from 1916 that lends additional backing to the bull shark theory. First, the general description of the creek attacker (like an "old weather-beaten board" with a white underbelly, about eight to nine feet long, and about three hundred pounds) matches a general color description of a bull shark and also matches the size of a mature bull. The eyewitness description could even match the proportions of Captain Cottrell's shark (seven feet, 230 pounds) when one takes eyewitness exaggeration into account. Additionally, the mannerisms displayed by the creek shark, before and during the attacks, coincide with Bigelow and Schroeder's description of the bull shark as heavy, slow swimming, and capable of swift movement when attacking.

A final area of evidence that strengthens the bull shark theory is derived from the fact that 1916 was an unprecedented year for the capture and sighting of the sandbar shark. The sandbar shark is implicated in five unprovoked attacks as listed in the ISAF. The authors of *Shadows in the Sea,* like most authorities, discount any direct involvement of the sandbar shark (also know as the brown shark, ground shark, and blue-nosed shark) in the actual attacks, but speculated whether its unusual abundance may have influenced an in-shore migration of the great white. Whites do commonly consume smaller sharks, but an equally common link exists (at least in tropical waters) between the bull shark and the sandbar shark. As discussed in chapter 5, the bull shark is notable among all sharks for its inclination to feed on the young sharks of other species, particularly the young of its near relative, the sandbar shark. The bays of New York and New Jersey were said to be alive with sharks in this particular summer, and unconfirmed accounts even reported that the Matawan Creek was visited by sharks that made the Wyckoff dock area a spawning hole. One doesn't have to get too imaginative to create a scenario in which a dangerous, northward-straggling male bull shark (or several) was drawn close to shore toward sandbar shark spawning grounds and happened to find his way toward the bays of Sandy Hook, Raritan, and smaller Keyport. The shark could have taken a familiar or unintimidating "freshwater turn" up the Matawan Creek, which ended in tragedy for Lester Stillwell, Stanley Fisher, and Joseph Dunn.

If we go along with the latest trend to implicate the bull shark in any gruesome attack that is not the obvious work of a white shark or a tiger shark, it might even be easy to consider the bull responsible for all five New Jersey attacks. In central and northern California, attacks are invariably attributed to great whites; in Hawaii, the tiger shark is often the named killer. South Africa and Australia have a mix of culprits, and Florida sees a mix of bull shark—or tiger shark—designated fatal assaults. At Florida's northern east coast, the shark "bite" capital of the world, the minor attacks are commonly attributed to smaller inshore species. New Jersey, however, with its classic four seasons, including a warm temperate, subtropical, and tropiclike summer, has migratory conditions that could include all large comers.

Just as there are many theories behind the New Jersey attacks, there are an equal number of writers and researchers attempting to answer the questions. In 1986, Reader's Digest published a comprehensive shark volume called *Sharks: Silent Hunters of the Deep,* which included a full chapter on the 1916 attacks, entitled "Death in New Jersey." The authors cited the pieces of human bone found in the Raritan Bay white, but they also mentioned that the link to a specific victim(s) in the New Jersey sequence could not be determined. They reported, "When the wounds on Stanley Fisher's leg were examined and measured it was found that the distance between teeth on opposite sides of the shark's jaws was about 14 inches. Comparisons seemed to indicate that a shark with jaws that size was probably longer than the shark witnesses reported seeing in the creek." Reader's Digest also moved in "bullish" fashion toward a more logical explanation, stating:

> The puzzle is that great white and tiger sharks rarely venture into fresh water. The one shark that is commonly found in fresh water, however, is the bull shark. Bull sharks have been found in the Mekong River [Vietnam], the Zambesi [Africa], the Mississippi [United States], as well as in Lake Nicaragua in Central America as far as 2,600 miles from the mouth of the Amazon.

In 1997, Mary Batten's popular *Shark Attack Almanac* also raised the controversy regarding the link between the dead 1916 attack victims and the human remains in the Schleisser white shark. Batten suggested a retrospective means by which the puzzle *could have been* solved. She asserted that DNA typing would have conclusively given the answer. Batten, too, found the "more likely culprit" (at least for the Matawan Creek) to be a bull shark.

Thomas Allen, one of the authors of *Shadows in the Sea*, wrote *The Shark Almanac* in 1999. Although in the *Almanac* he does not attempt to revisit the 1916 attack story with any earth-shattering new assessment, he does summarize the major point, which was made in the earlier approach. After discussing Edwin Thorne's findings regarding the ballooning sandbar shark population, Allen stated:

> An increase in the number of brown (sandbar) sharks in New York waters would have had no direct connection with the New Jersey attacks. But the increase did raise the question of whether a population explosion in indigenous sharks somehow had brought about the appearance of the white shark.

When *Great White Shark* was published in 1991, it was immediately obvious that Richard Ellis and John McCosker had put together a most thorough examination of "the most terrifying creature of the ocean." In his discussion of the 1916 attacks, Ellis included some of the fresh facts and memorabilia I had uncovered in my own research. However, this time Ellis implicated the bull shark species not only in the Matawan Creek attacks but in *all* of the 1916 attacks. Ellis even included a bull shark section in a book that was dedicated to the great white.

In *Great White Shark,* Ellis wrote:

> Probably the most celebrated shark attacks in American history took place in the summer of 1916, and although a white shark was then believed to have been the culprit, a closer examination of the evidence seems to implicate one or more bull sharks. Still, these attacks, and their aftermath, seemed to set the tone for press reportage and public response in later years, and they serve as an apt prelude to those attacks that are known to have been committed by white sharks.

In reference to the damning human bone fragments found in the Raritan Bay white shark, Ellis "made no bones" about saying that their species identification is still inconclusive, regardless of Dr. Lucas's assessment. He pointed out that Lucas was making some other very questionable statements of "fact" during this period. Ellis is not fully convinced that the Schleisser shark *was* a white shark, and he mentioned that "only the flesh and bones in its stomach tie it to the New Jersey attacks, and none too conclusively." He also reported that "whatever evidence there was is long gone. . . ." Ellis does admit, however, that whites are known to inhabit the mid-Atlantic bight (Raritan Bay is an arm of that bight), but that no evidence has been demonstrated to show their inclination to enter freshwater there or anywhere else in the world. The only large shark to have such a propensity, Ellis stressed, is the bull shark. In the final analysis, Ellis determined that the earlier Beach Haven and Spring Lake attacks where the work of one or another species of carcharhinid sharks, but that the nature and location of the Matawan attacks very likely implicates one bull shark.

In the midst of his discussion of the 1916 attacks and their probable culprit(s), Ellis provided a rather fearsome shark photograph from the July 16, 1916, *New York American* newspaper. The shark was allegedly caught on July 15 and the photo caption that ran at the time read:

> Vicious looking head of a blue-nose shark caught yesterday off Belford, New Jersey. The body measured 9 1/4 feet long. Belford is less than ten miles from the mouth of Matawan Creek, in which a man and a boy were killed and another boy seriously mangled by a man-eating shark. It is possible that this picture shows the head of the monster that has been terrorizing the New Jersey Shore for ten days past.

The caption that is provided by Ellis in *Great White Shark* stated, "On July 15, 1916, shark hunters caught this "blue-nose shark" [probably a bull shark] off Belford, New Jersey, during a frenzied hunt for the Jersey maneater." (The astute shark mystery analyst should take note that Captain Cottrell's captured shark was also identified as a "blue-nose.") In *Book of Sharks*, Ellis alluded to Cottrell's shark as the possible bull shark marauder, but in *Great White Shark*, perhaps after reviewing some evidence (i.e., John Cottrell's shark purchase and the element of flimsy creek fence netting) that was brought to light in *In Search of the "Jersey Man-Eater*," the Belford catch now seemed more culpable.

I must remind the readers that Richard Ellis has refined his theories to a subtle degree since his *Book of Sharks* as I have since *In Search of the "Jer-*

sey Man-Eater." Such modifications are the practice of every responsible and productive researcher. His basic premise remains the same, and it is to that premise and the original presentation he made (in his early work) that I direct some of my forthcoming comments. His remarks will be delivered in detail and the "subtle" refinements I have made in my theory will be presented as well. For that you will have to read on.

An examination of the evidence compiled by accomplished natural-history researchers like Richard Ellis and our growing knowledge of bull-shark behavior lends quite a bit of support to incriminate the bull shark(s) in the 1916 attacks. Theories are subject to questioning, particularly when new facts come to light. Despite the evidence in favor of the bull-shark theory, I believe much unappreciated evidence exists that prevents the bull-shark and the multiple-shark theory from being considered the most probable explanation. For this we must turn to ongoing research.

As with any sea creature that is not fully understood, much can be learned about its behavior through tracking and migratory patterns. Tagging programs that track the migration of sharks have provided a great deal of information on general long-term seasonal movements, but tagging is of limited help if one is hoping to determine the day-to-day movements of a shark. The taggers would be extremely lucky if a fisherman were to catch a shark within a day or two of tagging and then report on the tagged shark. Sharks that are captured the same day (and it does happen) are caught by the same tagging boat, after being, essentially, followed by the same persistent shark. It has been in only the last twenty years, therefore, that acoustic telemetry and other automated electronic homing devices have been convenient enough, accurate enough, and utilized enough to determine more about such movements.

The assertion that it is improbable for one shark to travel long distances in a relatively short period of time is not consistent with what we are now learning about many species. A 1982 paper entitled "Temperature and Activities of a White Shark, *Carcharodon carchar-*

ias" (written by F. G. Carey, J. W. Kanwisher, O. Brazier, G. Gabriel-son, J. G. Casey, and H. L. Pratt) described an experiment using acoustic telemetry to track a three-and-a-half-day swim of a large white shark from Montauk Point, New York, to Hudson Canyon, a distance of over 100 miles. In a similar experiment, a large white shark was tracked for two and a half days and covered 120 miles. Its average cruising speed was two miles per hour. The fastest sustained trip for a shortfin mako was made in three months over a 1,500-mile range, averaging 17.6 miles per day. The results of these tracking experiments, therefore, indicate that it is neither impossible nor improbable for one shark (at least not a mackerel shark—whites, makos, and porbeagles) to cover the distance measured for the time frame given in the 1916 attacks [that is, of forty-five miles in five days (Beach Haven to Spring Lake) and another thirty miles in six days (Spring Lake to Matawan)].

It has also been proposed that the wire netting placed up along the bridges to close off the Matawan Creek shortly after the creek attacks would have provided the Raritan Bay white shark with the perfect alibi. The creek was reportedly "penned-up with three rows of nets" at one point. In other words, it has been suggested that since the creek was closed off, the white shark caught in the Raritan Bay on July 14 could not have been responsible for the attacks because the creek attacker was still penned in. A New York newspaper even dramatized the breakout claiming that the net was forced down by the rampaging killer some three or four days *after* the tragedies and *after* Schleisser caught his shark. However, it is nowhere near certain that this netting was set up securely or thoroughly. In fact, reports say this netting was set up very hurriedly during the postattack hours. The *Asbury Park Press,* a newspaper much more local to the area, wrote of a reporter riding down the creek in a motorboat as early as the morning of the July 13 (one day after the attacks) and finding the initial wire netting to have been placed on the Matawan Bridge after having fallen down, being driven down, or being taken down.

Finally, the capture of the seven-foot, 230-pound shark, allegedly caught near the mouth of Matawan Creek by Captain Cottrell and his son-in-law, Richard Lee (this was the shark Ellis originally considered to be a bull shark and the only shark actually captured in the creek), does not conclusively prove that a bull shark was the only species capable of surviving in the creek. For one thing, many or most newspaper publications were fiercely apt to report these shark catches much closer to the creek than what was the case. This episode was no different. In fact, Cottrell's shark was caught while he and Lee were approaching the mouth of the creek. In other words, it was caught in Keyport Bay. Also, if Bill Burlew's account of John Cottrell's purchase is accurate, the shark may not have been caught in the vicinity at all. Additionally, although unsubstantiated, many Matawan residents reported that the creek was "alive" with many small sharks in 1916 and, in 1936, the *Matawan Journal* even reported that other sharks were seen in the creek from time to time in years following the tragedies. Two days after the attacks at Matawan, the owner of the Fisher bag factory, John Fisher, was "positive that he saw several sharks in the creek where the attacks had occurred." As mentioned earlier, I have even read a 1916 *Matawan Journal* report of small hammerheads being spotted in the creek. Incidentally, small hammerhead sharks have been found in Texas some two hundred miles up a river that connects to the Gulf of Mexico. In *The Sharks of North American Waters,* Jose Castro reveals that the smalleye hammerhead (*Sphyrna tudes*) does, in fact, enter freshwater.

Regardless of the reliability of reports of other sharks making their way into the creek, local fishermen identified the Cottrell shark as a ground shark (likely a sandbar) or a blue-nose/diamond-toothed shark (coloquial for sandbar shark, according to the *Brooklyn Museum Quarterly*, October 1916). This same coloquial name is given to the shark captured in Belford, New Jersey, which in *Great White Shark* Ellis says is "probably" a bull. Bull sharks do not have a bluish tinge.

Henry W. Fowler, author of the *Lamprey and Sharks of New Jersey*, points out that all photographs he has seen of sharks caught near the creek in 1916 appeared to depict brown sharks and that he believed a shark of that species was responsible for the attacks. John Nichols did not consider the sandbar shark a major potential culprit because they were always common in the vicinity. Nichols also doubted whether they had the "mechanical vigor" to do the reported damage. D. H. Davies and John E. Randall do list the brown shark as a dangerous shark, and as Thomas Allen cites, it is potentially dangerous because of its abundance but never implicated in an attack.

One issue that Fowler's and Ellis's assessment of photographic evidence brings to light is the very similar general physical characteristics of the bull shark and the sandbar shark. They are often confused even today. The bull has a somewhat more blunt-rounded nose, a shorter snout length, and a less superiorly (upward) tapered, slightly less angled, and shorter dorsal fin. The bull also lacks an interdorsal ridge and has more heavily serrated teeth. Beyond these differences, however, the major difference is that the bull reaches a rare but recorded larger size of ten feet, while the sandbar reaches only seven and a half feet.

As we know, from a behavioral standpoint, the bull is much more aggressive. One other similarity, often not emphasized, between the bull and sandbar is that both are commonly found in shallow inshore brackish areas such as harbors and bays and near wharves. The sandbar shark has even been known to venture into shallow water that rarely covers its body.

I would not be as confident in the sandbar shark's predisposition or capability to enter brackish water if it were not for an unexpected telephone call I received in 1996, a day after the debut of the Discovery Channel program *Legends of Killer Sharks*, for which I provided technical assistance and was interviewed. The program presented a fifteen-minute segment on the New Jersey attacks. The phone call was from Mrs. Lois Shafto, a woman in her early seventies who lived four miles south and inland of Asbury Park, in an area called Shark River Hills. The Shark River is a progressively narrowing waterway that extends approximately ten miles inland from its mouth at the Shark River estuary. The press of 1916 conceded that this river was once occupied by sharks during the 1600s but since then it had not regularly seen any shark inhabitants. In 1917, it was even suggested that the name Shark River be changed because, after all the shark hype of the previous summer, the name of the river might give a negative connotation to the area. The widest point of the river is a baylike region that bulges approximately two miles in any direction and opens to the sea another two miles distant through a standard, somewhat narrow, inlet. Mrs. Shafto's family roots dated back to the early founding of Monmouth County; indeed, an area just north of the river was named Shafto Corners. Mrs. Shafto called me to voice some dissatisfaction with the Discovery Channel presentation because I had failed to discuss Shark River.

In all the years my family has entered and exited the Shark River inlet with our fishing and dive boats, we never saw or heard of any sharks entering the inlet or following the river. Personally, I always wondered why they ever named that body of water Shark River inlet. How about flounder inlet or snapper inlet? I'd thought. Dr. Robert Patterson had offered me his explanation for the name. He said it was a body of water traditionally used as a sanctuary from rough seas during storms, and relieved captains originally dubbed it Shirk River because they would "shirk" storms there. Yet another explanation for the name came from over a century ago, when shacks filled with panhandlers and vagabonds used to line both sides of the inlet portion. Those inhabitants were said to evade traditional jobs and in essence they would "shirk" their duties. Either one of these explanations seemed quite logical to me, and, taking into consideration progressive phonetic change, I could see how Shark River might have evolved. At least those were my thoughts before the Shafto call.

Mrs. Shafto told me that, as a little girl, she would come home from school during the late spring and play by the small dock in front of her house, which was situated at the upper reaches of the river. "And the sharks would swim right around us. Most of them were small and never bothered us," she said. She recalled that they were called brown (sandbar) sharks. In one short phone call I had potentially solved the mystery of Shark River and received some fairly reliable testimony that the sandbar shark, like the bull, does not have much of a problem with low salinity. What Mrs. Shafto was likely describing was the once-common tendency of sandbar sharks, in the early part of this century and before, to penetrate the bays of New York and New Jersey for spawning purposes, even negotiating the relatively small bodies of water like the back reaches of Shark River. It had just never occurred to me that the sandbar sharks of Shark River would negotiate the fairly long narrow inlet and continue upriver. Why those sharks are no longer present there in such numbers is a mystery, but it may relate to intense commercial fishing, waterfront pollution, motorboat traffic, or some unidentified factors. It should be noted that, although it appears that the the sandbar has an obvious tendency to dwell in estuaries, Jose Castro, author of *Sharks of North American Waters,* does not believe that it penetrates into freshwater.

The photograph of the "blue-nose" provided by the *New York American* and reprinted in *Great White Shark* (and identified as "probably a bull shark"), presents some puzzling elements and perhaps another bit of controversy. The photograph of the shark, I first assumed, depicted either a sandbar shark or a bull shark. If the shark was truly 9.25 feet, as the caption describes, then I would venture to say it is a bull shark because of its length. The designation of "blue-nose" to Captain Cottrell's shark might somehow relate to its general coloration, as supported by Castro's *Sharks of North American Waters,* in which the author notes a bluish-gray variety for the sandbar shark. The Cottrell shark was also referred to as a ground shark, which is another common name for the sandbar or carcharhinid species in general. Then again, perhaps both the Cottrell shark and the Belford shark were the same species. Perhaps they were both bull sharks. Or perhaps they were a sandbar shark and a dusky shark or both dusky sharks. When you think about it long enough, it all becomes very "perhaps."

When examining the 1916 newspaper accounts for hard facts, one should closely scrutinize the statements, especially when they are not direct quotations. A rule of thumb would be to double the amount of scrutiny that you would use even by today's standards. Media ethics and accuracy policies were not as stringent at that time, and one must also consider that this case made vultures out of investigative writers. Suddenly, a person who had never seen a live fish before was responsible for covering a story about a large shark caught near a bay somewhere in New Jersey. The word "blue" in blue-nose could have even been a phonetically generated error derived from the designation blunt-nose, which could be used for something that looked like a bull shark. Then again, an even simpler explanation is that the writer simply needed a name for the ominously pictured shark, so he chose one he had heard thrown around at the dock. It is also curious that the shark in question was reported to be 9.25 feet long, which would be one of the largest inshore catches during the entire hunt, yet the camera angle is at the front end (missing central upper teeth and all), which discounts any opportunity for scale against which to compare the shark's proportions.

On Tuesday, July 11, 1916, Paul Tarnow, an established bayshore pound fisherman, caught a large shark off Belford, New Jersey, which was pictured in *Scientific American* a few weeks later. The newspaper writers who covered Tarnow's catch were faced with a dilemma because they were about to run the shot as it related to the July 1 and July 6 attacks. When the catastrophe occurred at Matawan on July 12, they were faced with losing some of the punch that would have gone with speculating on the guilt of the Tarnow shark. Their solution was to wait to run the picture a few days *after* the Matawan attacks. The *Keyport Recorder* did not hide this fact, but perhaps the *New York American* did. Efforts to pin blame on the Tarnow shark became trickier, however, when the Schleisser white shark was caught in the same general location as the Tarnow shark (but three days later). The Tarnow shark was reported as eight feet long in the *Evening Mail* newspaper, nine feet long by the Keyport paper, and was nine and a half feet long according to the Philadelphia papers. The shark had badly cut up the pound nets in the Keyport Bay and was caught only 120 feet from shore. The shark reportedly weighed 250 to 325 pounds. Tarnow did come upon another shark in his nets on July 15, but this time he kept the shark in a net corral for viewing. The second shark, which was "even bigger than the first," was not available for photos (dead) until a few days later. Tarnow's corraled fish brought him a gate profit of $300 dollars for one day. Interestingly, the first shark caught by Tarnow had twelve to fourteen small sharks (eighteen to twenty-four inches long) in its stomach when cut open. I'm fairly certain the first shark caught, the one featured in many papers on July 15, was the same shark as the one pictured in the *New York American* on July 16, because I possess reports of the sale of the actual shark to that paper for $8.

The picture in *Scientific American* reveals several fishermen around the tail of a shark and the shark hanging from a large hook. The shark appears to be approximately seven to nine feet in length, and my first assumption was, again, that it looked to be either a bull shark or a sandbar shark. In the *Scientific American* photo, the Tarnow shark is hanging on a hook and has a dorsal fin that appears much more like that of a sandbar than a bull, and appears to have an interdorsal ridge (an elevation between the two dorsal fins). It also appears to be female from the size of its pelvic fins. In other words, signs point to its being the standard large female brown shark of the region. The shark head in the *New York American* photograph is said to come from Belford. However, my belief has been that it is one and the same with Tarnow's shark, which would mean that it was caught *before* the Matawan Creek attacks.

Considering the subtle differences between bull sharks and sandbar sharks, distinguishing them is not easy when less-than-optimal black-and-white photographs and limited angles are all that are provided. My inspection of the *New York American* picture revealed that the shark's snout was indistinguishable (to my eye) from a sandbar and, although the shape of the teeth appeared more bull-like, the number of upper teeth appeared more fitting for a sandbar shark. Admittedly, the *New York American* shark was reported to be nine and a quarter feet, which makes it more likely a bull. However, the Tarnow shark featured in *Scientific American* looks to be between seven and nine feet, so it could be that the length in the *New York American* was reported inaccurately.

Upon closer inspection of the hanging Tarnow shark, however, I spotted the relationship between the dorsal fin placement and the pectoral fins. The shark's dorsal fin was clearly farther back along the spine than in either the bull or the sandbar shark. The bull shark and sandbar shark have dorsal fins that originate at about the midpoint of the pectoral fins (the pectoral axil). What shark was this? The carcharhinid shark that has a dorsal fin placement at that location and that matches the general description as well as distribution territory and size for this shark is the dusky shark. The dusky shark even has more of a tendency to possess a blue tinge to its grayish skin than does the sandbar. It grows to a longer size (over eleven feet) than the bull shark. My identification of the *New York American* photograph, the Tarnow shark, and Captain Cottrell's shark must at this point include, or even place at the top of the list, the dusky shark. The fact is, blue is a variable coloration for the sandbar and the dusky shark, not for the bull shark. It becomes notable to recall that in Edwin Thorne's extensive local shark population studies he caught only two types of sharks: the abundant sandbar shark and the *dusky* shark. As far as documented attacks go, the dusky has been implicated in about the same number of human assaults as the sandbar. The dusky is one of the more abundant large sharks found in New Jersey.

Stanley Fisher's death certificate. Cause of death reads: "Bitten by shark."

CHAPTER 8

The Wounds Tell a Tale

I f a daring theorist attempts to pin any one of the 1916 attacks on the Cottrell shark, the Tarnow shark, or the Schleisser shark, the question of human remains and wound characteristics must be evaluated. During the last half of the summer of 1916, the practice of inspecting a shark's stomach contents became commonplace and potentially profitable if human remains could be identified (assuming the shark was registered at a bounty-friendly town). Much has yet to be learned about the digestive processes of sharks, but it seems that human flesh or most any food item will remain undigested for long periods. The metabolic rate and digestive rate of elasmobranchs are generally lower than for bony fishes, and much lower than for mammals. Apparently, a unique intestinal valve increases surface area of the absorptive intestine, thereby decreasing needed length of the intestinal tract. The nature of this intestinal apparatus, however, requires a much longer transit time for food. The active shortfin mako takes two days to digest an average-size meal while the sandbar, blue, and lemon sharks take three to four days. By contrast, the more sedentary nurse shark takes six days or more to digest its meal. Some experts believe that the great white, depending on its activity level, body size, and meal size, can go as long as a month between feedings.

The most dramatic example of a shark's digestive timing comes from the infamous "Shark Arm Murder Case." The account, which is detailed in Dr. Victor M. Coppleson's book *Shark Attack,* tells the story of an Australian fisherman who hauled in a fourteen-foot tiger shark in 1935. The shark ended up in the Coogee Aquarium, but after several shaky days of survival, it eventually regurgitated its stomach contents in front of a crowd of fourteen interested onlookers. The spectators saw the distinct remains of a rat, a seabird, and a human arm emerge from the digestive scum. The arm was very well preserved and had a rope tied around its wrist with a nautical knot and a tattoo on the forearm of two boxers going nose to nose. One of the tattooed pugilists had blue trunks and one red trunks. Coppleson, a famous Sydney surgeon and shark-attack specialist, was called in and found that the arm was severed by some artificial but unprofessional means and was not dismembered by any shark. After briefly considering a medical school anatomy prank, the police finally discovered that the arm belonged to a pool-hall operator named James Smith. The identification was accomplished by chemical peel applied to the fingertips for a successful print match. Smith was reported missing several weeks before and was known to have been involved with some shady characters. Ultimately, Smith's roommate from a local boardinghouse was charged with murder. The man implicated was Patrick Brady, and only days later a second murder occurred. Before the second victim was killed, he'd insisted that Brady had been out to murder him because of his knowledge of Brady's guilt in the Smith case. The charges did not stick against Brady. The first trial ended with a hung jury, and in the second he was acquitted. Brady died in 1965, declaring his innocence to the end. Whether he was truly guilty or not, the fact remains that Smith's arm was sheltered in the stomach of the tiger shark for at least eight days (the time the tiger spent in the aquarium) and possibly up to eighteen days (the time since Smith was said to have disappeared). Since the arm was in such good condition, it is possible to conclude that the shark's digestive juices even had some preservative qualities (if the arm had spent ten days exposed to seawater it probably would have looked much worse).

The fact that a bird, rat, and intact human arm were regurgitated by the tiger shark might tell us a great deal about a shark's digestive qualities. I have seen dissections in which birds with intact feathers are removed from sand tiger sharks, while the same shark will possess only the skeletal remains of a dolphin. This would seem to indicate that the chemical digestive process of some sharks does a more effective job digesting fatty items than bony or muscular items. Perhaps some sharks are simply better endowed to assimilate and break down fat. As already touched on, pinnipeds (small marine mammals), whales, and dolphins are major food sources for certain large sharks, and these animals possess a major fat component. By comparison, a human limb is a much more muscular and bony item. In the 1916 case, quite a bit of flesh was removed from each victim. Bruder and Stillwell had significant bone loss as well. Therefore, any guilty shark caught in the days following the attacks would still have possessed some form of human remains in its stomach.

One promising aspect about wound analysis, even as it relates to this antique case, is that it offers concrete evidence in an attempt to identify an involved species. Speculation or debate about the type of wound is not really at issue in the 1916 case even without photographic evidence. The eyewitness physicians provided their reports outside the unpredictable sphere of press reporting. Their descriptions are detailed enough to gain a fairly clear view of the injuries and make it possible to compare such wounds to similar shark-inflicted trauma of photo-documented modern attacks.

The very serious upper thigh wound to Charles Vansant is fairly simple to envision and describe. The upper leg was denuded of flesh at the back (postero-lateral) and outside (lateral) portion of the thigh from hip to knee, down to the bone. The femoral artery was severed

Charles Vansant

and the right thigh included a gash. This wound did not contribute to his death. Vansant's left thigh wound essentially comprised a cleanly extracted/carved wide elliptical piece of multiple layers of tissue. The skin, the fascia, the vastus lateralis, and the vastus rectus of the quadricep group and the hamstring were all partially removed. The nerves, arteries, and veins (neorovascular bundle) with the all-important femoral artery were severed. From the wound alone, one could surmise that this was a large shark (at least six feet in length), and likely had at least an upper row of serrated, bladelike teeth. Such teeth belong to sharks of the carcharhinid (Requiem) family (which includes tiger sharks, bulls, and sandbars) as well as the great white of the Lamnidae group (mackerel sharks).

There are varying accounts of the attack pattern of the shark assaulting Vansant, and many of the details depend on at what elevation and distance an onlooker was standing. Some people were on the beach near a lifeguard area, and others were walking along a slightly elevated boardwalk. One account, at first glance, is so fantastic that it almost comes across as pure fabrication. It is the version that tells of the shark biting down on Vansant's leg and remaining affixed to his thigh until Vansant and the shark are dragged into only eighteen inches of water. How a witness would come up with such a specific depth is yet another curious feature of 1916 data. Perhaps it was an estimate established by someone like swimmer Alexander Ott, or perhaps it was someone stressing the fact that it was a shallow depth less than two feet deep. This "shark-affixed version" was even part of the account that Jackie Ott conveyed to me. However, such a sensational detail actually has equivalents in other shark attack narratives, and it may ultimately provide valuable information. A strikingly similar incident occurred in New South Wales, Australia, on February 26, 1966. In that attack, a thirteen-year-old boy named Raymond Short had begun to tread water just outside the surf. He was first seized on the left thigh, then the lower part of the right leg. The boy punched and kicked the shark as hard as he could, but the shark would not let go. A lifeguard raced out to Short as the shark warning bell rang. Other lifeguards joined in the rescue but, despite Raymond's cries, none of them were aware that the shark was still attached to him. At the youngster's insistence, one of the guards ran his hand down the affected leg through the murky, churned surf and felt the shark. As they made it into shallower water, the body of the beast was revealed. A surfer attempted to bash the shark off with blows from his board, but the jaws stayed clenched. On the beach the shark was eventually forced to release its dry-docked victim. In the end, the front and back of the boy's left thigh had multiple lacerations. His hands were also badly lacerated from defensive blows. The right leg, however, suffered the worst injury. There, the calf musculature was virtually gone, and the shinbone (tibia) bore teeth marks from end to end. The shark, which died on the beach in full view of hundreds of beachgoers, was originally identified by assorted names such as a mako, a whaler, and a gray nurse shark, another example of the inaccuracy a shark culprit inspires in the heat of the incident. The trained eye of a professional easily determined the species sitting on the beach. It turned out to be an eight-foot three-inch, immature, female great white shark.

The wounds to Charles Bruder and Lester Stillwell were easily the most gruesome work of the Jersey man-eater. In the case of Bruder, the soft tissue injuries were described as a cleanly cut gash ("a piece cut out the size of an apple or a fist") along with the sliced bone of the region of both distal (lower leg) extremities, which displayed slash-type lacerations to the remaining soft tissue. There was also a portion of tissue gouged out above the left knee. Such wound characteristics in South African Natal coast attacks (which normally don't involve many white sharks) are seen as distinctive traits of the carcharhinid sharks (bull, tiger, etc.). The late D. H. Davies, in *About Sharks and Shark Attack,* saw the presence of clean-cut wounds on Natal victims as positive proof of an attack by a shark with flattened, triangular, serrated teeth. To Davies this was synonymous with the M.O. of a bull shark. To me, it seemed as

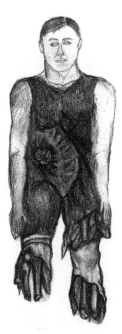

Charles Bruder

Lester Stillwell

if Davies attributed the bull shark label to any Natal case where clean-cut wounds were present. Later, his colleagues at the Oceanographic Institute of South Africa, utilizing modern knowledge of tooth fragments, population studies, and attack records, reidentified many of Davies's bull shark designations as the work of an assortment of car charhinids. Today, however, the Natal Sharks Board still considers the bull shark the most notorious attacker on the Natal coast.

Regardless of the debate in attaching a specific species to clean-cut, singular, large wounds, the shark responsible for such an injury would most likely possess at least an upper row of serrated teeth. The jaws, which are equipped with such continuously aligned cutting teeth, are invariably powerful and utilize a crushing capability that performs complete limb amputations (despite what Dr. Lucas said) and slices away flesh with such efficiency that massive hemorrhage normally ensues. In contrast, deep penetrating, spikelike wounds are more the work of sharks with nonserrated, knifelike teeth (e.g., odontaspids, makos, etc.) and pose a greater infectious problem.

Interestingly, Charles Bruder's wounds were not all cleanly edged. Apparently, the Spring Lake victim's legs, what remained of them, were torn and clawed as though something had hacked them, and not bitten them with a clean, sharp, razorlike bite. The appearance of the end portions was inconsistent with the other cleanly gouged portions on Bruder's body. According to Victor Coppleson, in *Shark Attack!*, these specific wound features were once used by 1916 theorists to suggest that Bruder had not been attacked by a shark at all, but rather a sea turtle or a huge mackerel. In any event, the apparent combination of clean-cut and tear-type wounds should be considered potentially significant in light of the forthcoming analysis of the wounds inflicted at Matawan.

Of all the 1916 victims, it was probably Lester Stillwell who generated the most interest as far as what damage had been done to the body. The two-day time span when Lester was unaccounted for created great interest and hopeless concern for the condition of his body. He was considered the only victim that the Jersey man-eater was able to take possession of. At first, it was assumed that Lester was fully consumed. Later, after the body was found, the analysts wondered whether the shark had been frightened away from Lester's corpse. No matter what happened between the shark and Lester after the boy was drowned, it seems as if the boy was completely overcome from the instant he was attacked. This scenario is quite different than the other attacks. Charles Vansant, for example, was quickly rescued and did offer a "man-size" version of resistance to the shark. Charles Bruder was sufficiently mangled, but even without his legs he seemed spirited enough to fight off the marauding shark. Stanley Fisher put up a very vicious fight as well and received a good amount of immediate assistance from men with slamming boat oars, etc. Joseph Dunn was literally pulled from the throat of the shark by his brother and a friend. Young Lester, however, was overtaken rather swiftly, and with no adult creek-side assistance, he likely perished rapidly from the combined affects of his wounds, shock, and asphyxiation. Did Lester's corpse give us a glimpse at the unhindered destructive potential of the Jersey man-eater?

Lester was more or less struck at all quadrants of his body. In view of the "bite and spit" hypothesis, and the assumption that sharks are not supposed to attempt consumption of man, let alone continue to "nibble" on him, the injuries to Stillwell are quite illuminating. Stillwell's body showed torn flesh and bone in the area of the left ankle (said to be "torn off"), his left thigh was mangled from hip to knee, the right hip (said to be torn off), the abdomen was torn open with intestines significantly torn out and revealed, the right chest muscle was torn away, and the left shoulder was torn away as well. It was not clear whether the shoulder injury included bones such as the clavicle, the acromium, and the humeral head, or simply the deltoid muscle alone. As mentioned in the narrative description, his face was untouched.

One thing that the Stillwell attack wounds make clear is that this was not a "bite and spit" execution. The shark initially latched on to Stillwell while the boy was lying virtually supine on the surface. His friends had heard him say, "Hey fellas, watch me float." Almost any of the wounds could represent that initial strike, and perhaps the others took place during the "fight" prior to the boy's death and sometime afterward. Lester was obviously not consumed, and his limb bones did not suffer the type of jagged-ended amputations that Bruder experienced. Because of the distribution, depth, and severity of the injuries, however, the creek shark most definitely sampled and chomped on Lester's corpse multiple times. The wounds he suffered were not caused by glancing, open-mouthed, inadvertent contact. The wounds also make the rumored Stanley Fisher statement about seizing Lester's body from the gnawing shark at the bottom seem credible.

Of the known attack patterns (i.e., "hit and run," "bump and bite," and "sneak"), what kind was involved with Lester Stillwell? The shark certainly did not "run" after it attacked Lester, unless it merely attacked, dropped the boy, then returned when Fisher, Smith, and Burlew were searching. The shark's tail struck Albert O'Hara shortly before the strike on Stillwell, but beyond that, there was no distinct "bump" or warning. The boys did, however, say that they mistook the shark for an "old black weather-beaten board." The shark (perceived to be a board) was probably half submerged (thought to be water-logged). Regardless of the possibility that it was spotted before its surge, the shark's attack was likely a variation on a "sneak" attack because of the severity of the biting that followed. In fact, I would assume that each of the 1916 attacks was more a "sneak" version than anything else because of the absence of a clear bump/warning and because of the quantity of body tissue removed. In other words, the shark showed some deliberate intent, or premeditation, to consume the prey item. In the cases of Stillwell and Bruder, the shark turned on the victim multiple times. In the case of Bruder, the shark likely went at him three or four times, and in the case of Stillwell, five times.

Stanley Fisher appears to have received one severe bite to the right thigh. His wound was descriptively similar to that of Vansant except for the fact that Vansant's wound was on the left leg and his injury was more posterior (toward the back of the leg). Unlike Lester Stillwell, Stanley was lucky (but not lucky enough) to have a crew of spectators nearby. The two men in the rowboat persuaded the shark to let go, as did Stanley. Stanley was also fortunate that a doctor had been summoned for Lester before Stanley was even attacked. With Stillwell, on the other hand, the shark had a generous amount of time and opportunity.

Even though Fisher's encounter with the shark seemed momentary, almost every piece of muscle and tissue from his thigh, from hip to two inches above the knee, was removed. The only portion remaining was the medial (inner) third. With the reports of blood spurting skyward, Fisher's femoral artery was likely severed. Recall that when Fisher first reached down to determine the extent of damage to his leg, he said, "Oh my God." The femur bone surface (periosteum) was exposed, but it was not penetrated, only scratched. About ten pounds

Stanley Fisher

of flesh was estimated lost, and the remaining tissue appeared to have been racked with dull knives. The extent of the wound (end to end) was clear. The physician who examined Fisher's wound in Matawan estimated it to measure eighteen inches, but it was later determined to span fourteen inches. It is possible that Stanley Fisher, only a few feet from assistance and possibly without arterial destruction at the femoral triangle, did not suffer the same rapid volume of blood loss that Vansant suffered. Both men expired after approximately the same period of time, but Vansant never regained consciousness after getting to the beach. Vansant probably lost great quantities of blood in the water and plunged into shock earlier.

Donald R. Nelson, a researcher at California State University, Long Beach, has classified general shark-attack behavior according to the situation (i.e., element of provocation). Unprovoked attacks are generally those in which a shark considers a person a prey item and the person has done nothing especially stimulating to attract the shark. An unprovoked attack with distress stimuli usually involves a shark that is motivated to attack because of splashing. Perhaps there is also blood in the water, and the shark may transcend its usual predatory behavior patterns. A third type involves provocation with physical contact, which occurs when a person mistakes a shark for a smiling dolphin and attempts to take a ride or simply play with it or torment it. The final form of attack, from a provocation viewpoint, involves an inadvertent or unexpected encounter between shark and man in which the shark feels threatened in some way and the person unknowingly fails to heed an aggressive/warning gesture. Here, the shark strikes and swims away. The reader may contemplate and correlate these discrete attack situations to the 1916 attacks, but it would be safe to say that one of the first two scenarios would apply to each of the attacks. Do not forget, however, that the Matawan Creek could have represented a threatening/unfamiliar situation to the potentially wayward shark (though perhaps not unfamiliar to a bull shark). Some newspaper witnesses even claimed the Wyckoff dock area was a spawning hole. In such cases, perhaps the fourth version, involving territoriality and defensive behavior, is appropriate. If it were a spawning hole for a female bullshark, her drive to feed would be uninhibited.

Joseph Dunn, the only survivor, experienced a severe wound to the lower leg as opposed to the upper portion (above the knee) of the extremity as seen in Vansant and Fisher. Charles Bruder obviously received distal lower extremity injuries but to a much more catastrophic degree than Dunn. Dunn appears to be lucky for two main reasons. Probably most important, his brother and Jerry Hourihan were at his side in an instant to pull him out of the shark's mouth. According to Dunn, "It felt as if I was going down the shark's throat." The other rea-

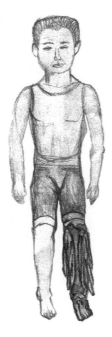

Joseph Dunn

son is likely the position at which his leg was clenched by the shark. Dunn's wound was described as ribbonlike lacerations to the front (ventral aspect) and side of the left leg from knee to ankle and severe lacerations to the calf. The major arteries and tendons were said to be intact. Some of the smaller bones of Joe Dunn's leg were said to have been penetrated as if they had been drilled by a sharp instrument.

To better understand the way in which such wounds were delivered to Joseph Dunn, a quick review of a shark's dentition would be helpful. Large dangerous sharks generally have three kinds of teeth. Sand sharks, mako sharks, and porbeagles have stiletto-shaped teeth. Other sharks are equipped with broadly serrated triangular teeth, such as in the upper and lower jaw of the white shark and the upper jaw of many carcharhinid sharks (including the bull shark, sandbar, and dusky). The teeth of the lower jaw of the carcharhinid sharks are narrower than their upper jaw and vary in serration from heavy to fine. The more pointed, narrow, finely serrated types of teeth, as in the lower jaws of many carcharhinids, are mainly for grasping prey. The heavily serrated teeth are made for cutting since they allow for slicing pieces of flesh when the shark's head moves violently from side to side. The tiger shark, also a carcharhinid, is an exception to the triangular-upper-row/narrower-lower-row configuration. The tiger shark has a partially serrated upper and lower row of teeth, but the teeth have a diagnostic notch, almost like a button hook.

Joseph Dunn had fairly distinctive wound characteristics on the shin side versus the calf side. From a physician's point of view, I would immediately guess that such a pattern was influenced by the fact that the front and back surfaces of the lower leg are quite different in design. The calf is a fairly fleshy mass made up of muscle (mainly gastrocnemius/soleus) and underlying nerves and blood vessels that exit behind the knee at the popliteal fossa. The end (distal) portion of the calf muscle ends with the Achilles tendon, which inserts into the heel (calcaneus) bone. Joseph's foot was also said to be severely lacerated in ribbonlike fashion. On the outside of the knee (the lateral side), the peroneus longus muscle, the peroneal nerve, and the narrow shaft fibula reside. Moving toward the front center (shin/anterior midline), we find the firm tibia bone running from knee to ankle. The very front (ventral/anterior) portion of the tibia is one of the few places in the body where there is no muscle between the skin and bone. Such a presentation makes it exquisitely tender to impact (like those shin kicks). Keeping all of the anatomical review in perspective, why would Dunn have ribbonlike ventral lacerations and severe lacerations to the calf muscle, without major arterial or tendinous compromise?

Let us compound the question posed above with the fact that many of Stillwell's wound characteristics (torn muscle and flesh and unfractured bone) appear to come from jaws that perhaps bear the long, pointed, smooth-edged, stiletto-type teeth. Such wound types and teeth are typical of sharks that tend to tear at the flesh of their prey, e.g., the mako shark (*Isurus oxyrynchus*), and the sand tiger (*Eugompho-*

dus taurus), etc. The great whites, bull sharks, and tiger sharks, on the other hand, commonly perform a cleanly cut operation with their sharp, serrated, triangular teeth.

Regardless of some jagged edges or the "racked" appearance of Fisher's thigh wound, he and Vansant had experienced the removal of a large amount of flesh from a localized site. Such a wound is classic for the serrated-toothed sharks. The length of Fisher's thigh wound (fourteen inches across) and the length described for Vansant's wound also point to a shark with serrated (at least upper) teeth. As I reported earlier, *Sharks: Silent Hunters of the Deep* asserted that a shark with a jaw width of fourteen inches would be larger than nine feet (longer than Schleisser's shark, which was seven and a half feet, and longer than witnesses estimated for the creek shark). A shark whose jaws measure fourteen inches in width would probably have a length over ten feet. In *In Search of the "Jersey Man-Eater,"* I explained, quite clearly and credibly I thought, that the wounds to Vansant and Fisher were not the result of a single clenching bite/snap, but "obviously" a raking of the shark's teeth on the limb of the victim. This raking is a prototypical, violent, side-to-side motion. Fisher's wound was not a perfect excision because Stanley was far from a willing victim. The shark was allegedly even forced to "spin him around and take him down twice." Dr. Perry Gilbert of the Mote Marine Laboratory explains that with the triangular serrated teeth of its upper jaw, a shark can easily cut deep into flesh and, with the aid of shaking, rotary, and lateral movements, can remove large chunks of flesh with the quick, knifelike fashion of a single bite. The so-called "torn" tissue could merely have been the result of openmouthed jerking movements caused by both the attacking shark and the frantic victim. These characteristics may also be associated with the lower jaw of a carcharhinid whose tooth edge is more finely serrated. Most of these possibilities would somewhat point away from incriminating a white shark whose upper and lower jaws are distinctly triangular and heavily serrated. A 1975 published report, however, submitted by members of the

Oceanographic Institute of South Africa, offers information on teeth studies that makes the exclusion of the white as a culprit impossible. This report stated that though the upper and lower teeth of adult great whites (greater than ten feet in length) are triangular and serrated, some of the teeth of younger specimens (e.g., Schleisser's white shark) resemble those of the mako, the sand tiger, and the porbeagle (*Lamna nasus*). Some of these immature white shark teeth may be smooth-edged and lanceolate. Also, the fact that great whites have relatively wide-spaced teeth (wider than most carcharhinids of this size class, including the bull), could explain the "dull-knived" groove-type lacerations on Fisher's thigh.

Shortly after the release of *In Search of the "Jersey Man-Eater,"* Richard Ellis and I traded lengthy single-spaced letters to debate particular issues relating to statements of fact or significance in the 1916 case. In reference to my point on the possible variability on the Schleisser shark's triangular-tooth composition, Ellis keenly observed that E. W. Gudger described the purported Jersey man-eater jaws (hanging in the Manhattan fish shop) to have broadly triangular saw-edged teeth. I made no excuses for what Gudger said, and I did not dispute the point or even suggest that Gudger's fish shop jaws might not even be the jaws of Schleisser's fish. I didn't even propose that perhaps *some* of this white shark's teeth were less heavily serrated and less broadly triangular than Gudger had described, nor did I suggest that my observations about the wound characteristics might be flawed. I just let the point stand, that is, until this time.

The fact is, even adult white sharks, or at least many of them, do not have mirror-matching triangular, heavily serrated teeth in their upper and lower jaws. If you look closely, even at the monster depicted on the front cover of *Great White Shark,* you will see that many of the lower-jaw teeth have a narrower, more pointed apex than the upper teeth. In other words, it is not uncommon, at least from my observation, to find less broadly triangular lower-jaw teeth in white sharks, despite the observation by Gudger that they are technically triangular.

Using a formula that Chuck Stillwell of the Narragansett National Marine Fisheries Service provided me, I was able to calculate the approximate jaw width of the Schleisser shark by examining the jaw-width to body-length ratio for juvenile white sharks. The seven-and-a-half-foot white shark caught by Michael Schleisser in Raritan Bay (some reasonably reliable sources of the day reported the shark to be eight and a half feet in length as well), would have had a jaw width of nine and a half to ten inches. Eye-balling the Schleisser shark in photos and noting his remark about the shark being "able to fit a man's head inside when open" pretty much confirms such a measurement. Recalling that Michael Schleisser was a Barnum & Bailey Circus animal trainer, he would probably have used the lion tamer's method: he would put his head straight into a lion's mouth and not turn it sideways. My head, of assumed average size, measures eight inches from side to side and 9.5 inches front (nose) to back. Therefore, if you attached a nine-and-a-half to ten-inch-wide jaw to the fourteen-inch thigh wound of Fisher, you can see how lateral movements of two to two and a quarter inches to either side would have produced the wound in question.

The resultant lower leg wound seen in Joseph Dunn does present a potentially interesting scenario as it relates to the position of his leg in the shark's jaws. Obviously, the fact that Joseph was tugged out of the shark's mouth by his brother and a friend had something to do with the fact that he survived and did not receive a more severe injury or multiple bites. But, if this were the same shark that attacked Fisher and Vansant, why did it not remove a clean section of flesh? Additionally, if this were the shark that sliced or crunched right through Charles Bruder's lower legs, why did he not do the same to Joseph's left leg? A look at the hotel doctors' examination of Bruder can shed some light on the controversy. They reported:

> On Thursday July 6th, 1916, between two and three o'clock in the afternoon, we were called to the beach to attend a bather,

said to have been bitten by a shark. On arriving we found a man, Charles Bruder by name, about 28 years-old, in a surf-boat on the beach. He was in a state of collapse and died in a few moments. On examination, we found that both legs were missing; bitten or broken off about 4 inches above the ankles and a large cut above the left knee. The flesh [was] torn off the right leg from just below the knee to where the bone was bitten or broken off, leaving the bones protruding without any flesh. A piece of flesh bitten out of right side below the ribs, also showing tooth marks.

Certainly the possibility that these attacks are not the work of the same shark in every case (and not even the same shark in the individual creek incidents) has to be considered. It could also be surmised that even the same individual shark may not inflict the same wounds on each of its victims or prey items. The Matawan Creek shark had enough time and opportunity to cut deeply (deeper) into the large bones of Stillwell, but it did so little bone damage and lacked wide tissue extraction that the appointed medical examiner felt that the damage was done by a small shark. Dr. Lucas, in reviewing the wound determinations from Colonel Schauffler, believed that the shark involved in the Bruder attack was also "not a very large one." While we are looking at bone penetration, we must also consider, if it were the work of a single shark, why did it not sever the legs of Vansant and Fisher?

When comparing the wounds of Vansant and Fisher to those of Bruder, one should take into consideration a particular anatomical feature. The Vansant and Fisher wounds involved the upper thigh muscles and the underlying femur bone. The femur is much thicker and stronger than the combined tibia and fibula (lower leg bones), and although a shark of this presumed kind (a large, dangerous one) might be able to sever such a bone, I am not familiar with too many instances when a shark ever has. The likely hindrance here is when the shark encounters the femur surface, it may simply realize that it has not bitten into a thickly blubbered marine mammal. The best the shark will end up with in this case would be equivalent to what you and I end up with when we chew on barbecued ribs. It may not be reaching to assume that the shark has some type of expectation of the consistency of the prey item it has targeted. In other words, when it pursues a large fish, it expects a fish taste and texture, and when it pursues man, it expects either a sea mammal's makeup or at least that of a sea turtle (if you accept the current mistaken-identity theories). Stewart Springer and others have long emphasized the rather deliberate process a shark undertakes prior to full-fledge feeding. From an earlier chapter, you will recall that many sharks use their teeth or mouths to examine or test the prey consistency. In this case, the shark may be turned off enough by the prey item and not bother to follow through to the bone or with general consumption. In Bruder's case, he may have suffered the "apple-size" abdominal wound as a test bite. The shark, detecting a somewhat acceptable fleshy-fat prey item (Bruder was stocky and not thin), "turned on" its predatory aggression and returned its aggression on Bruder's lower legs with bites. Even if these subsequent bites were still test bites, they were forceful enough to sever the lower leg bones. Most important, the shark did not (could not?) sever the upper, wider, stronger tibial (shinbone) of Bruder on either leg, but rather at the lower, thinner end. I'm sorry to use human dietary behavior (like Lucas and Murphy in April 1916) to explain the shark's methods in the Bruder attack, but the effective removal of flesh, without severing the bone at the same time, does appear to be similar to a person gnawing chicken meat off a drumstick.

The situation with Joseph Dunn could be a case in point as it relates to the "turn off spit" aspect of the "bite and spit" hypothesis. When the shark bit down on Joseph's lower leg, its upper jaw, presumably equipped with broader triangular serrated teeth, immediately encountered the obviously bony tibial (shin) region while its lower jaw (with pointier, finely serrated teeth) encountered the fleshy but shallow muscular tissue. The limb did not present in a way that the shark could perform lateral rotary tissue excision. To this shark, which probably just confronted two unappetizing victims up the creek, sensitivity to determine the appropriateness/worthiness of prey might have even been heightened. Some accounts speak of the shark coming from below and behind, and others tell of a "test-warning" bump or scrape. The fact that Dunn's leg was not snapped even puzzled the physicians in 1916. According to the *New Brunswick Times,* the hospital staff mentioned that the boy had the "frame of a midget" and they assumed "his plucky fight against the monster frightened it away before it could complete its destructive purpose." In the end, the shark tentatively clenched the upper shin and calf and caused the resultant wounds as it raked downward on the release. Joseph's tibialis anterior muscle (lateral to the shinbone) was cut to ribbons, and the calf muscle was "severely lacerated." The double row of lower jaw teeth, which did not possess the regularly arranged teeth pattern of the upper teeth, did not sever the major deep-seated arteries, nor did they cut through the Achilles tendon. In the case of Lester Stillwell, recall that: "The boy's left ankle was chewed off, his left thigh was mangled from hip to knee, his left abdominal region was ruptured and the intestines herniated and torn open. The intestines were nearly all torn out. The right hip, right chest muscle, left shoulder, as well as several fleshy areas of the body were all eaten away and the flesh between his right hip and thigh were mangled. His face was untouched." The right hip, left shoulder, and possible ankle wound may have included significant bone loss, but I would not consider them large, long-bone assaults. That the shark seemed almost to avoid severing the larger bones of Stillwell supports the contention that the bony texture (which you would expect in eleven- and twelve-year-old boys) of the younger victims contributed to the feeding "turn off" in those cases. The wounds on Stillwell could have even been the result of the shark's tentative (nonfeeding) initial test bites (i.e., the shark attacked Stillwell, drowned him, released him, and did not continue to chew on the body). I am in no way suggesting that the wounds to these boys were minor, nor am I saying that amputation-like wounds to children have not occurred on many occasions in the past.

Stewart Springer, who has done a great deal of interesting work on predatory feeding behavior in sharks, may have also hit on a feature of feeding inhibition that could relate to the Matawan attacks. Springer points out that the disposition of sharks actually to feed may be variously inhibited by circumstances or conditions in this environment. In other words, a shark is less likely to feed in an unfamiliar environment (e.g., Matawan Creek). As alluded to in regard to Donald Nelson's general attack classifications, hunger may not have been a motivation in any or all of the creek attacks. The shark could have been acting out of fear, agitation, aggression or territoriality. Territoriality or possessiveness may be a fair argument in the case of the attack on Fisher if the story of the Stillwell body retrieval has foundation in truth.

One other reason that Fisher may have been attacked more viciously was the fact that blood was in the water. I am well aware that experiments dispute the contention that blood arouses a shark to attack or feed even though it is a substance that sharks are naturally drawn to. I am also aware that sharks are supposedly less interested in mammalian blood as compared to fish blood. I have a few thoughts as to why such arguments don't stand up to scrutiny under all circumstances. Most notably, sharks, including the bull shark, the tiger shark, and especially the white shark, hold marine mammals as sought-after prey items. The special features of the white shark's stealthy-approach tactics are made to order in its quest to snatch smarter, more maneuverable marine mammals such as seals and sea lions. Admittedly, there are some differences in the makeup of marine mammal blood and human blood, especially as it relates to hemoglobin and albumin concentrations, but my major point is that sharks have evolved a supersensitive mechanism to track down prey based on blood content. Even though that blood alone does not drive a shark into an aggressive state, I most certainly believe it becomes interested in examining a situation where blood is present. Additionally and, perhaps, for our purposes, most important, the factor of competition and stimuli have to be weighed. In a shark frenzy or heightened-arousal situation during which, for example, a dead fish is being torn apart by a few sharks, the instant the sharks sense competition, the frantic biting will ensue. Imagine Stanley Fisher, Arthur Smith, and Red Burlew making multiple dives amid bloodred water for the tattered corpse of Stillwell. Whether Fisher got to the body or not, this activity could have certainly given the shark the impression that some type of competitive feeding pursuit was occurring.

Regardless of my comments on blood and sharks, potential rescuers of victims who must swim amid bloodred water should not be overly concerned about being attacked by the lone shark in the vicinity. The facts are fairly clear that rescuers are only very rarely injured by the shark that has already attacked a person. Apparently, a shark will not redirect its aggression to another prey item very quickly. Or possibly it has already decided that the entire event is a mistake. If the shark does not mistake the rescuer or the situation as competitive or threatening, then it generally will not attack again. The shark in the Fisher incident may have even become possessive over the boy, as was the case in the bizarre grappling-hook incident in Australia where a white shark became incensed as Coast Guard members were attempting to recover a drowning victim. The shark in the Australian case barely allowed the crew to remove the body from the water.

One scenario that has not been discussed is the timing of Stillwell's specific injuries. Young Lester was so bony, we might assume he was immediately released by the shark in an instantaneous "bite and spit." The initial strike was enough to asphyxiate the boy, but when did he acquire the multiple widespread wounds? Did the chewing take place when Fisher and the rescue team were diving, in some sense of competitive chomping, and/or after Stanley fought for the boy's corpse?

In reference to the severity of wounds and the commentary on those wounds in the Matawan Creek, one other topic of some interest is worth considering. Although thorough comparative jaw strength studies of the white shark, bull shark, and other dangerous sharks do not exist, some assumptions could be drawn from the size of the prospective creek attackers. If the creek killer were the Schleisser white shark, we would be talking about wounds inflicted by a juvenile shark (with heavy and fine serrated teeth) in alien surroundings on unnatural prey. For the bull shark scenario, we should consider the eyewitnesses' length estimates and the lengths of the captured local sharks. Such a length would come within a range of seven to nine and a quarter feet. According to Castro's *Sharks of North American Waters,* the average size of an adult male bull shark is 7.5 feet and a female, 7.8 feet. In *In Search of the "Jersey Man-Eater,"* I assume that a mature adult bull shark would possess more power, more destructive efficiency (even if it were not actually feeding) than a juvenile white shark, and probably would have inflicted more deeply avulsive wounds upon each victim in the creek than was the case. I am not completely convinced of such an assumption, but in view of the fact that freshwater rivers are semifamiliar surroundings for the bull, I would guess that a very large adult bull shark would be at its fearsome best in such a situation.

Evaluating the 1916 attack wounds as they compare to wounds of modern-day victims and the associated outcomes is also quite interesting. In view of the inferior form of first aid, transportation, and treatment of eighty-five years ago, the severity and nature of the victims' wounds must be closely examined, as opposed to a simple look at the resultant fate of the victim. Forty years ago, G. D. Campbell and David Davies wrote a book on the pathology and treatment of shark-attack wounds, and in the publication a classification system was established. That classification system is still quite relevant even by today's standards. Their grading of shark-inflicted injuries is helpful in our research, and it is defined as follows:

Grade I injuries involve major damage to the femoral artery (in the area of the femoral triangle) or where severing of multiple arteries takes place. The victim usually succumbs within minutes after the attack and the outcome is always fatal.

Grade II injuries involve less severe arterial or abdominal damage, when the victim usually survives if correct treatment is administered on the beach.

Grade III wounds are the most common and are the major form of wounds encountered in the relatively frequent Florida attacks. These involve minor arterial, abdominal, or limb damage when the victim always survives if treated correctly on the beach.

Of the 1916 victims, Vansant and Fisher had very similar wounds, with both men experiencing femoral artery damage. The fact that they both survived for greater than a few minutes is attributed to the very speedy tourniquet action by Alexander Ott in Beach Haven and Dr. Reynolds at Matawan. These vicious wounds were not Grade I injuries because the arterial damage did not likely occur at the femoral triangle. This anatomical triangle is generally located in the inner, upper third of the front portion of the thigh. One reason why this location is so threatening is the fact that it is so high up on the thigh that using a tourniquet or providing manual pressure is almost futile. The severed artery may even retract under the ligament to make it even more inaccessible. The major reason that blood loss is so rapid and deadly when this portion of the artery is involved relates to the bore and pressure at that point. The artery is larger there than in the lower portion of the thigh and therefore the rate of blood flow (loss) is extreme. Poiseuille's Law shows that when blood flow (Q), pressure differential (P), and radius (which here is r^4) are divided by the product of blood viscosity (n) and the length (1) of the vessel, the size (diameter or radius) of the vessel plays the greatest role in determining the rate of blood flow through the vessel. In the early 1990s in South Africa, a sixteen-year-old girl was surfing when she sustained a massive upper leg wound in the area of the femoral triangle. She was one of the few for whom fate was reversed and she survived. When her blood filled the water seconds after the attack, a fellow surfer got off his board and went to her aid. The man placed his fingers and hand right up and firmly onto the spurting artery, practically at her groin. The girl required an amputation of the limb, but she survived to work in the fashion industry and serves as a role model to other victims of disabling diseases and accidents.

In the case of Lester Stillwell, the shark did not dismember the boy nor did it seem to devour great chunks of tissue with each bite. Obviously, a slight-framed eleven-year-old boy does not possess meaty sections of pliable tissue to extract, but the shark in this case did cause a very serious combination of injuries. Let us not forget that the water was bright red from the trauma applied to Stillwell. His shoulder and the opposite chest muscle were torn away. With those portions of anatomy avulsed, major arteries like the axillary artery could have been severed. One ankle area was also torn away, and the abdomen was ruptured with the intestines exposed. The intestinal injuries could have even damaged the mesenteric arteries. The attending medical examiner did report the contributory cause of death as "Hemorrhage & Shock." Stillwell's wounds, therefore, would have to be considered Grade I, and even without major initial catastrophic blood loss, the trauma alone would certainly throw the boy into devastating shock.

Charles Bruder's injuries were without question Grade I in quality because of the multiple arteries that were severed and the catastrophic blood loss that ensued. He died within minutes in the bottom of the lifeguard boat.

Joseph Dunn, the only survivor, was undoubtedly lucky in several ways. He was fortunate to have assistance at arm's length, and he did not experience major arterial damage. If the shark had gone only an inch or so deeper, the multiple lower leg arteries (peroneal and tibial) behind the knee would have been involved. His wounds were Grade III.

In retrospect, Vansant and Fisher's chances for survival would have been increased if they had received standard modern first-aid and transportation, including state-of-the-art shock treatment (saline infusion, morphine, and oxygen) in the field. Even if Vansant and Fisher had survived, they would have required amputations above the knee.

If one looks into enough shark-attack cases and listens to the comments of the victims, one will notice a perplexing phenomena relating to the victim's pain perception. Attack victims frequently greatly underestimate the severity of their wounds and even report a minimal amount of initial pain perceived. Quite often, recovering victims will say, "I really didn't feel pain," or in the case of the Dunn boy, "it was like a big vice pressing on me" or "like a tug from a big pair of scissors." At Matawan also, Stanley Fisher did not even know he was severely injured until he made his way to shore. You will recall that one of the examining physicians in Matawan hypothesized that sharks emit an anesthetic toxin with their bites. We now know that sharks do not release any type of numbing poisons, but some experts suggest that the first bite of an attack produces massive nerve damage or somehow numbs the victim's sensation of pain. Such a mechanism is similar to the numbing phenomenon that shock victims of standard traumatic accidents experience. Another explanation, however, would be to link this response to our body's own mechanism of pain management under stressful circumstances. The neurochemical mechanism at play here is not fully understood but may include neurotransmitters or other hormones that are commonly released under extreme circumstances. Those chemicals include epinephrine, serotonin, endorphins, enkephalins, etc. Such chemical-release patterns are well established in soldiers and athletes, and the actual existence of this neural mechanism has been generally accepted. The specific pathway involved and the exact reason for such a system continue to be areas of debate. Today, pain-management physicians are utilizing multiple approaches to the treatment of chronic and acute pain, and such approaches include the standard opioid narcotics, low-dose antidepressants (which stimulate the stress/analgesic neurotransmitters mentioned), and spinal cord stimulators. In *Principles of Neural Science,* Eric Kandel and James Schwartz suggest that ". . . there are many behavioral situations in which an organism's normal reaction to pain could prove disadvantageous—for instance, during predation. . . . Perhaps stress induces analgesia in the most extreme, life-threatening situations." I also learned from a conversation with dentist and pain physiologist Ken Hargreaves of the National Institutes of Health, that distraction can be an effective analgesic. Possibly the "distraction" of the attack—with its suddenness, unknown origin, unpredictability, and life-threatening quality or predatory nature—can inhibit or delay the victim's perception of pain.

This pain-free first few minutes of an attack is not specific to the 1916 attacks. Ashley Walker, a twelve-year-old from Chillicothe, Ohio, who was visiting Pineland, North Carolina, can attest to that. Walker had never seen the ocean until she went for a swim on the Outer Banks in July 2000. The girl stated, "I thought someone was grabbing me. It really didn't hurt." Initially, she felt something brush against her, and then a tug. She didn't realize how badly she was hurt until she reached shore and saw a nine-inch crescent-shaped gash on her right calf. The wound required more than three-hundred stitches. She made a complete recovery.

One other mechanism of action that may contribute to this initial deadening of pain in the attack victim relates to a bombardment of stimuli. The well-known gate theory presented by K. Melzack and P. D. Wall describes how certain touch receptors, besides carrying tactile (touch) function, will also stimulate a "gate" (substantia gelatinosa) in the spinal cord to inhibit pain. Such a feature can be better understood with a simple analogy. A certain road (the pain pathway) is used for a delivery truck (the pain sensation equivalent) to drop off milk. If the truck gets stuck in traffic (the overwhelming or synchronous stimulus), it may not reach its destination (the brain). This concept is utilized in the physical therapist's pain-reducing modality of transcutaneous electrical neural stimulation (TENS). This mode of action has even been suggested for acupuncture. Speaking of acupuncture, some theorists have suggested that the initial attack bite might trigger an "acupuncture-like" initial effect. Acupuncture has been around for more than three thousand years, and its Western medical understanding points to the induction of relief of muscle spasm with the insertion of hair-thin needles, triggered release of natural analgesic substances (naloxone, a narcotic inhibitor, has even inhibited the acupuncture pain-relieving effect), triggered release of ACTH (the brain's anterior pituitary hormone that stimulates cortisol, a cortisone-like hormone), the gate theory effect, and even hypnosis. However, regardless of the potential crossover into viewing a shark's teeth/bite as an acupuncture-like phenomena, I don't think the topic is worth further examination or discussion.

David Livingston, the famous Scottish missionary who explored Africa in the early 1800s, may have been referring to the same concept of traumatically induced analgesia when he wrote of an experience with an attacking lion. His entry notation in *Missionary Travels* is a fitting and eloquent conclusion to the discussion of the attack victims' strange perception of pain. Livingston reminds us that even though we are to have "dominion" over all the animals on the planet, we still have an inherent fear of predators, just as our cave-dwelling counterparts feared the fierce jaws of carnivorous beasts. He wrote:

> I heard a shout . . . and looking half around I saw a lion . . . he
> caught my shoulder as he sprung . . . he shook me as a terrier

does a rat. The shock produced a stupor . . . it caused a sort of dreaminess . . . there was no sense of pain nor feeling of terror, though I was quite conscious of all that was happening. This particular state is probably produced in all animals killed by carnivora . . . and is a merciful provision by our benevolent creator for lessening the pain of death.

Advertisement of a viewing of the alleged Jersey Man-Eater.

CHAPTER **9**

Explaining the Inexplicable

The ultimate question, of course, regarding the 1916 attacks is: Why did they happen at all? It is a given fact that the 1916 attacks were special in many ways, but it is additionally perplexing to consider that the Jersey Shore does not see a substantial frequency of shark attacks. New Jersey is not a "red triangle" or "a shark-attack capital" or deserving of any of the terms that describe areas which see a seasonal swell of attacks.

The East Coast of the United States has always had an abundance of sharks, but the inshore shark concentration around New York and New Jersey in 1916 was certainly unusual. The presumed 1916 "shark invasion," therefore, stands as a topic that must be confronted and can be most illuminating in regard to its influence on the string of five attacks.

To attempt to solve the riddle behind the apparent mass migration of sharks in 1916, ascertaining fact from fiction is critical. In many cases the apparent shark sightings and shark catches at that time may have been simply the product of more active shark fishing, overactive imaginations, or both. During the shark-frenzied period that followed the five attacks, the front pages and sports sections of newspapers were filled with stories of tremendous shark trophies. Rich folk, poor folk, pretty women, famous figures, and anyone else who could find a sturdy rod and a guide were out to get a face-to-face look at the "devil from down in Jersey."

New Jersey, New York, and other Atlantic coastal states were seeing the new sport and obsession of shark fishing take off. The sighting craze was also contagious. In late July, for example, in the Chesapeake Bay, Maryland, near the mouth of the Patapsco River, several large sharks were reported. (Could these have been bull sharks?) Despite the fanatical behavior toward sharks, it is quite difficult to dispute the presence of an abundance of large sandbar sharks in the New Jersey/New York area, at least one large juvenile white shark, likely large dusky sharks, sand tiger sharks, and perhaps even bull sharks. Many other poorly identified large sharks were also caught by shoreline fisherman and bay fishermen. The photo that depicts a *Field & Stream* writer standing on the beach after reeling in a relatively large shark speaks volumes about how unusual the beach bathing situation really was. Three reliable contemporaries of the 1916 attacks, George Burlew, Bill Burlew, and Johnson Cartan, each living in the vicinity of the Keyport Bay, all corroborated the reports that there was a shark explosion. The testimony of Edwin Thorne and his seventeen-year population study speaks for itself. We should also consider, with some degree of scrutiny, that steamship captains were reporting enormous numbers of shark schools heading toward New Jersey or schooling just off the coast. One report identified some fifty-six schools offshore, with scores of fish per school.

There is evidence, however, of a relative shark invasion in 1915. John Nichols, Bill Burlew, and the testimony revealed in some Coast Guard records make this point. A July 17, 1916, letter from the Seaside Park, New Jersey, keeper reported that in the fall of 1915, fishermen along the beach were so inundated by sharks that three or four channel bass would be chomped in two each night by sharks right off the beach.

As much as I disagree with some of the 1916 shark-related state-ments of Frederic Lucas, his theory of a shark year probably holds some water. Lucas theorized that some unappreciated factors had caused an abrupt and sudden appearance of numerous sharks just as some years see hoards of "butterflies, moths, army worms, jellyfishes, western grasshoppers, or northern lemmings." As early as the Vansant attack, Ocean County bay fishermen, clammers, and oystermen were reporting many more sharks than in most previous years. What we are speaking of is at least a sudden recognition of large sharks in the New Jersey shoreline and bay regions. But what could be responsible for such an influx, and how does such an influx relate to those first twelve days of July? The problem with the simple shark year theory is that 1915 and 1917 were probably shark years as well.

We do know that most sharks take periodic migrations based on available food, reproductive cycles, and environmental features. These migrations can include short distances over a day-to-day time frame or seasonal travels over great distances (even transatlantic cross-ings). On the basis of seeking food, some sharks simply vary their rou-tine by spending the daylight hours deeper, then ascending to the upper water levels at night. Some simply move from offshore to in-shore at night. Still others follow their prey along the prey's long distance migratory routes.

Reproductive cycles also play a key role, and perhaps in 1916 they were most important in drawing many sharks of one species to the general area. Some sharks gather for mating purposes in enormous schools composed of uniform size or sex. The fertile mothers will mi-grate to specific nursery areas, like New Jersey bays, and shed their eggs or give birth to pups. The nursery areas are chosen to allow for the most optimal chances of survival for the young, which often in-clude locations like coastal and estuarine waters. These "sanctuaries" provide abundant bait fish, crustaceans, and shelter from larger predators. Unfortunately, today, the coastal nursery areas are also the prime areas for habitat degradation. According to *The Shark Almanac,* when speaking of estuaries, "here is where navigation channels are built and maintained, where levees and marinas are developed, where sewers and factories discharge their wastes, where pesticides and fertilizers run off." Speaking of environmental impact on estuaries, on September 14, 2000, a hole in the hull of the Navy supply ship U.S.S. *Detroit* spilled thirty thousand gallons of diesel fuel into Raritan Bay. Despite the continual threats to the nursery regions, preliminary results of nursery studies of sandbar sharks show that shark distribution may be related to salinity and distribution of prey fishes, and juveniles seem to move in groups with the tides.

According to Jose Castro, environmental factors such as temperature, oxygen solubility, light, and oceanic currents affect the migrations of sharks. This combination or the specific contribution of these factors is not well understood, but clearly temperature is the most obvious and important environmental influence. It determines water density, oxygen solubility, and influences the metabolic rate of sharks. Since most sharks are cold-blooded and have body temperatures that correlate with the surrounding water, each species has a range determined by its metabolism and adaptive abilities. Sharks may even be more tolerant of colder temperatures than they are of warmer conditions. In Florida, for example, during abrupt cold snaps when water temperatures have dropped, bony fishes have been found dead in hoards, while sharks come swimming in to devour the dead fish. The June 24, 2000, Jersey Coast Shark Anglers shark-tournament program printed out a "shark catch" chart that relates water temperature and the species most likely to be caught at that temperature. The blue sharks, which are known to prefer cooler waters, are best caught between fifty-seven and fifty-eight degrees Fahrenheit, with catches dropping swiftly above sixty-three degrees. In 1980, Collier and Miller found great white attacks in Central and Northern California occur at sixty degrees or less. Makos, sandbar sharks, duskies, and tiger sharks appear to bite very well when the water temperature is in the low seventies (seventy-one to seventy-two degrees, to be exact). (Could this information relate to Coppleson and Davies's "temperature attack belts," where attacks are most likely to occur above sixty-eight to seventy degrees?) The hammerhead seems to prefer the upper seventies for more bait bites. These variations are the reason why some sharks will migrate in order to find the temperature range that is suitable for their tolerance or preference. Generally, such migrations involve moving northward and inshore during the summer and southward and offshore in the winter. The temperate areas of the world see the most dramatic migratory changes because of the wide variability in water temperature and the dramatic changes in conditions during the turn of seasons.

Sharks that are involved with a seasonal migration pattern will display the most rapid and steady rate of movement. Dusky sharks, for example, have been known to cover thirty-two nautical miles in one day during a weeklong southerly autumn migration. Other records have even documented an eighty-mile migration in one day.

Other environmental factors that have been assessed by tagging programs include current systems, annual rainfall, abundance of prey, intensity of fishing, and pollution. The U.S. government's first shark-tagging program began in 1962 in the Atlantic Ocean, directed by Jack Casey out of the Sandy Hook Marine Laboratory in New Jersey. To better understand certain migratory predispositions, one must first understand the three general categories in which sharks are placed. The *pelagic* species are those that range over broad stretches of ocean and generally occupy the deep offshore regions. Pelagic species include the oceanic whitetip, the blue shark, the porbeagle, and the mako shark. *Coastal pelagic* species are those confined to the continental shelves but have shown movements exceeding a thousand miles. These include the sandbar shark, the dusky, the tiger shark, the sand tiger shark, and the white shark. *Local* or *resident* species are those that spend most of their lives in one particular range of a few hundred miles or less. These local species include the nurse shark and the bull shark.

Tagging has given us astounding data. A sandbar shark tagged off Long Island, New York, was recaptured in Progreso, Mexico. Dusky sharks and mako sharks have been known to travel almost two thousand miles, and tiger sharks greater than eighteen hundred miles, with a known trip from New York to Costa Rica. Multiple recaptures have even suggested that blue sharks may make round-trip movements between North America and Europe (exceeding ten thousand miles) and sandbar sharks could easily travel the thirty-five hundred miles between Southern New England and Yucatan, Mexico.

The great white is one shark that has little compiled tagging data to analyze. Since the white shark is relatively rare everywhere in the world and shark fishermen rarely release one for tagging purposes, not much is known about its standard movements. The small amount of data that is known from tagging white sharks is actually confounding; it does not depict a predictable pattern. In other words, the shark is known to make fairly long-distance movements from one area to another, but many white sharks are caught in the same area in which they were initially tagged during opposite months.

In the inshore areas of the western North Atlantic off Long Island and New Jersey, shark tournaments have been held annually for more than twenty years. The tagging that is performed in those controlled situations has told us more about the white shark population in that vicinity than any other method. Out of 5,465 sharks caught in that region, 26 were white sharks. Most of these whites were juveniles, but very large whites have been caught in this mid-Atlantic bight region as well. The large white sharks are said to feed on whale carcasses off New Jersey, and there are filmed episodes of five or six individuals simultaneously lunging on top of a dead whale to chisel out chunks of blubber. Jack Casey et al have made some conclusions as to the patterns of distribution among the whites and have found that "white sharks are likely to occur singly or as scattered, unassociated individuals, over several square kilometers." The authors of *Great White Shark* wrote: "White sharks seem to be loners, and do not often associate in schools, or even pairs, as far as we know—until word of a whale carcass or shower of tuna chunks goes out."

As a volunteer member of New Jersey's Marine Mammal Stranding Center for more than fifteen years, I can attest to the fact that the New Jersey waters are populated by marine mammals and potential prey items for large white sharks. During late winter I have gone out on numerous occasions to assist in rescuing sick or injured seals that have beached themselves. Sometimes I've helped out three different seals within a ten-mile stretch within one day. These species are mainly the harbor seal, the harp seal, and the hooded seal. I've also encountered porpoises and dolphins of many kinds (dead and alive) and, during the summer, the bottlenose dolphin, which shows up in large schools on a regular basis. Some of these schools even develop a daily beat, where they will come in on certain beaches to feed or play in the morning and late afternoon. I have even helped rescue a seven-hundred pound bottlenose with a kidney infection from the now familiar Keyport Bay and in October 2000 I assisted in the National Marine Fisheries rescue of a mother bottlenose and her calf from the Shrewsbury River. Besides the pinnipeds and small whales, large leather-back sea turtles, loggerhead turtles, and Kemp-Ridley turtles show up battered by boat propellers or display swordfish longline hooks hanging from their very dead mouths. In other words, the white shark can find some tasty treats along the Jersey Shore, although nowhere near the density of inshore marine mammals that central and northern California, the Cape Seas of South Africa, or parts of Australia hold.

Any sudden appearance of sharks in any location that does not regularly see such abundance is probably not due to a reproductive population burst (of that shark species), but rather to a shift in distribution caused by some dramatic environmental change. According to Chuck Stillwell of the National Marine Fisheries Service in Narragansett, Rhode Island, seasonal changes due to weather patterns and other natural events might force the shark to alter its normal habits. There were several environmental factors of the summer of 1916 that could have influenced variation in shark behavioral patterns. June 1916 was a very warm month, breaking many heat records and encouraging many ocean bathers into the water (the number one factor to correlate with shark attack). The ocean water temperature in July 1916 may have actually been lower than average because of strong southerly winds and water currents. The ambient temperature in June 1916 hovered around eighty-five to ninety degrees and even ventured into the low nineties. The predominant June wind was northwest, which was very abnormal for that period of the season. That wind generally brings cool, clear seas. By the first week of July, the winds were steady out of the direct north, which brought cooler, more comfortable temperatures toward the region. On July 5, the wind was briefly out of the northeast, which brought even cooler air to the area but brought warmer water. As we know, the northerly winds seem to carry Gulf Stream waters close to land. The ocean temperature during the Vansant attack was a comfortable sixty-eight degrees, and right before the Bruder attack the water was in the low seventies. Just before Bruder had entered the water on the afternoon of July 6, the water temperature had taken a plunge into the mid- to low-sixties because of the southerly change in the wind. For the next week, the wind would remain out of the south with cool seas. The air temperature would remain in the high eighties to low nineties. The high temperatures and humidity promoted crop and vegetation growth, increased insect populations, polio transmission, and, perhaps, shark attacks.

One of the first years in which human refuse was piped into the central coastal waters of New Jersey was 1916. This process was being used at increased volumes while the seasonal population increase was swelling. May and June were unusually rainy months, perhaps influencing nutrient-rich runoff that thereby would influence small creatures of the food chain to flourish, and so on, up the ladder. It rained on consecutive days from June 3rd to the 11th and from the 16th to the 21st. The average number of rainy days was four above normal. The inshore salinity may even have been altered during the rainy period. Jack Casey speculates that the blue shark is actually deterred from entering inshore waters because of its reduced salinity. As mentioned, the blue is a pelagic species and prefers the much saltier offshore waters.

Many steamship captains reported an unusual westward shift of the warm Gulf Stream waters off the New Jersey Coast, but this was not necessarily proven, nor was an increase in offshore temperature documented.

For some reason, commercial fishermen were said to be especially guilty of disposing of fish entrails and unused fish parts close to shore during return trips. The findings of Victor M. Coppleson and G. D. Campbell seem supportive of some of these potentially relevant factors and the 1916 attacks, especially in regard to the Matawan Creek. Dr. Coppleson wrote:

> Recent heavy rain, debris, a swollen creek and an ocean, mixed in practically any proportions, comprise the almost perfect ingredients for a shark lure. The flood refuse washed into the ocean provides succulent food for the hungry marine mouths. It attracts a chain of fish and other sea creatures, beginning with minute sea organisms, each devouring the smaller ones in front. At the last link in the chain of survival are the ravenous sharks, waiting to pounce on the final victors. Many of the attacks on Australia's east coast and along the coast of Natal have occurred near the entrance of rivers, creeks, and lagoons in muddy water after heavy rain.

Finally, overfishing and depletion of menhaden stocks was mentioned by more than one observer in 1916. The menhaden population or availability was said to be at a twenty-five year low. In 2001, the National Marine Fisheries Service acted to better preserve the East Coast menhaden population. Menhaden is recognized as a staple for many larger fish species, and its recent demand as an omega-3 fatty acid source has raised concerns about overfishing.

In relation to murky, muddy waters and attacks, Tim Wallett has reported that though murky and turbid water may be coincidental with many Natal attacks, it does not appear that such conditions induce "anger" or viciousness in the sharks, or else many more attacks would be taking place. Some evidence does support, however, the correlation between diminished visibility and the chances of mistaken-identity attacks. The poor visibility may also influence a shark to make precipitous decisions because of the short amount of time between spotting a potential prey and snatching it before it dashes off.

The water temperature (the bathing areas along the Jersey Shore were in the range of sixty-six to sixty-eight degrees during the period of the attacks) does coincide with Coppleson's average critical range of sixty-eight to seventy degrees, as well as the shark-catching statistics and the sharks' apparent willingness to bite. Shark fishermen—professional and amateur alike—were doing a fine job with capturing sharks during that same period. The "critical temperature" is said to be "warm enough to put sharks in the mood and swimmers in the water." G. D. Campbell even thought that elevated water temperatures may produce a state of "petulance" in sharks, whereby they seek out prey regardless of their state of hunger. Campbell's assertion likely does not apply to the 1916 case since water temperatures were below seventy.

During late July 1916, government-hired vessels arrived at the conclusion that the Gulf Stream had not shifted as contended by some theorists. The first chart of the Gulf Stream was drawn in 1769 to the order of Benjamin Franklin to teach English mail packet ship captains how to make the same swift westward Atlantic passage as the experienced American fishermen. Franklin had heard about the Gulf Stream flow from a Nantucket skipper who used the current to expedite easterly travels and intentionally avoided it coming west (against the current). The Englishmen refused to take the advice of "simplistic colonial fishermen" and spent two weeks longer than necessary during trips from England to Boston. In 1916, a hydrographer for the U.S. Navy, Captain Thomas Snowden, theorized that cyclonic storms passing along the New Jersey coast influenced winds and produced a pressure wave that disturbed the normal Gulf Stream pattern. The normal pattern for the immense "river" of warm water is to come up from the southern U.S. shore waters (it comes closest to shore around Palm Beach, Florida) and flow northward and arc along the coastline. As it moves north, it moves a bit farther offshore and makes a right turn (as does the contour of the coast) and heads eastward around New England and Nova Scotia toward Europe. The Gulf Stream is about fifty miles wide and a strip of colder water, called the "cold wall" by sailors, separates it from the Atlantic coast. Jennifer Clark of the National Weather Service informed me that the phenomenon that Snowden described could occur, but using a formula based on depth and current flow (the one hundred fathom curve), she estimated that the major flow of the Gulf Stream would never come closer than about eighty miles from shore in New Jersey. One difference between the coast of Monmouth County, New Jersey, and Palm Beach County, Florida, is that the water depth off New Jersey's beaches takes an extremely gradual sloping course as you move offshore. In New Jersey, you would have to go several miles to get to a depth of one hundred feet and head to the assorted canyons some sixty miles offshore to achieve much greater depths. In Florida, the water drops to one hundred feet in many locations within a mile of the beach.

The fact that the major flow of the Gulf Stream did not likely change course during the summer of 1916 does not, however, mean that the Gulf Stream did not have at least some effect on the Jersey Shore during that time. Each year, much to the delight of New Jersey divers, the Gulf Stream drops off small tropical fish species, which have somehow hitched a ride and straggled into the many bays, shipwrecks, and jetty rocks by late August and September. Walt Campbell of the Ocean Products Center, National Oceanographic and Atmospheric Association (NOAA) in Baltimore, Maryland, informed me that even though the major flow of the Gulf Stream may not move in closer than eighty miles along the Jersey Shore, the huge river of water it encompasses is not completely uniform in shape (it may take on a "clover leaf" arrangement at times). Therefore, it is possible for certain weather conditions actually to break a small section of the Gulf Stream away. Campbell stated that this stray loop of warm, clear, nutrient-rich water (similar to the Gulf Stream composition) would be a potential lure and a good "drop-in diner" for many forms of marine life. He also claimed that he swam in such pockets of water near the shore in Florida, Georgia, and North Carolina. Could the Gulf Stream have dropped off a tropical terror such as a tiger shark or a bull shark? The renowned American artist Winslow Homer apparently thought the Gulf Stream was associated with some "sharky" inhabitants. His intriguing painting *The Gulf Stream* depicts a lone black sailor attempting to ride out a storm on a battered fishing sloop. The poor man's small boat is surrounded by three large, frenzied sharks.

Despite the seemingly unusual factors present during the summer of 1916, one is left with an unavoidable question: Since the conditions alluded to have probably been present many times over during the past eighty-five years, why haven't the extraordinary events of 1916 been repeated again, even once, in New Jersey or elsewhere? Could it be that the factors referred to are not as significant as other less apparent or unknown factors? Or, perhaps, the New Jersey attacks of 1916 are simply a one-time freak of nature. One unsettling thought is that similar occurrences are, and have been, only a moment (or a bite) away from happening, but, by some twist of good fortune or sheltered degree of publicity, less eventful outcomes transpired hundreds, if not thousands, of times over.

For the armchair theorist, the environmental influences offer a wide margin for attaching some fantastic possibilities to the cause of the attacks. For example, if sharks do commonly make the long-distance trips described and do react to forced unnatural and natural changes in their environments, could it be possible that the World War I North Sea bombing did chase dangerous sharks across the Atlantic? In 1916, some were even speculating that bombing practice in Cape Hatteras, North Carolina, drove sharks (like the bull shark) northward. The U.S. fishermen of 1916 were already convinced that the sand tiger shark was also know as the Spanish shark because the heavy cannonading during the Spanish-American War (1897) drove the shark away from its natural habitat and influenced their migration north ever since. And what about those German subs, the *Bremen* and *Deutschland?* These early U-boats must have had excessive vibrations, electrical emissions, and crude waste (both refuse and metabolic) disposal methods and could have easily served as an inadvertent attraction to some sharks. The *Deutschland* was even moving in a mysterious northerly course from Baltimore, Maryland, to Connecticut during the attack period. The New Jersey attacks, of course, took place in a progressively northward sequence as well. World War II Pacific sea disasters notoriously attracted sharks to the area of sinking ships and stricken sailors, and the stimulus of grinding metal and dispersing organic materials seemed to be an almost immediate lure (e.g., the infamous disaster of the World War II cruiser *Indianapolis* where 850 men survived a Japanese torpedo, only to have 500 of the survivors consumed by the sea and the packs of sharks).

The shark's ability to detect certain electrical or electromagnetic fields has been well documented and is assumed to relate to their ability to locate prey that possess a natural electric field and even for determination of directionality based on the earth's magnetism. Some commercial shark and sword fishermen are now even utilizing this known biologic attribute to lure the sharks to them from long distances. Vibratory factors are known to attract sharks from very long distances, and this fairly specific and convenient commercial electromagnetic device emits a field that is said to draw sharks from four miles or more. John McCosker, of the Steinhart Aquarium in San Francisco, believes that white sharks may have a higher sensitivity to electric fields than any other animal ever studied. He has also suggested this ultrasensitivity may be the reason why whites sometimes rake their teeth on metal boat propellers or shark-proof cages. These metal objects apparently possess their own electric fields. I wonder what the sharks would have thought of the primitive 1916 submarines. The sharks could have even initially confused the objects for whales. I, too, believe that white sharks have an extraordinary electromagnetic sense and likely do respond to the artificial metal objects because of electromagnetic field properties. The sharks could, however, simply be determining the texture and tensile qualities using their teeth for sensory purposes, much as humans would use their hands.

I am personally puzzled about one aspect in regard to the German U-boat situation. The U-boats' presence here was very controversial. They had always purported to be present in our ports for humanitarian-aid reasons, to assist their stricken fatherland with benign merchant opportunities. For instance, they were going to transport condensed milk back to Germany for the hungry babies. The United States and the government officials heatedly debated the subs' status as warships, and since these submarines were here allegedly on pure peace terms and just "happened by" when New Jersey and the East Coast were in an uproar over sharks, why didn't the sub captains at least make an offer to help us in our efforts against the sharks?

Regardless of one's conjecture about whether the submarines were able to attract sharks to shore, the Natal Sharks Board has developed a new electric shark repellant called the SharkPOD Diver Unit that takes advantage of the shark's ultrasensitivity to electric fields. The unit attaches to the scuba tank and the electrodes attach to a fin and a shoulder area. The distance between the electrodes generates a spherical field which is said to repel the sharks from one to seven meters away. The unit is advertised to be effective against the top three dangerous sharks encountered in nearshore waters: the great white, the bull, and the tiger shark. The literature claims that "tests against the great white have been extensive and the results convincing." Tests against the tiger shark and the bull shark have been limited but the results "have been good." Trials against various species are said to be ongoing. The unit is intended to be turned on for the duration of a dive and not simply when a shark is spotted. The manufacturers would probably prefer to deter an approach by a shark rather than to deter an oncoming attacker in the act. The promotional literature also mentions that it is "hoped that the SharkPOD will ultimately become standard equipment for any human exposed to the dangers of the ocean's top predators—the sharks." Personal defensive devices such as shotgun shell-loaded "bang sticks," CO_2 pressurized projectiles, and electric darts have also been employed.

A discussion of the environmental factors that provided us with a capsule of known circumstances during the summer of 1916 should also include a more detailed presentation on extuarine biology and the tidal marsh. Such a topic has an especially relevant connection with the influx or abundance of the sandbar sharks up and down the coastal bay regions of New Jersey during that summer.

Mudflats, with their accompanying marshes, are commonly found in estuaries—the regions where rivers and streams empty into the sea. Here, freshwater flowing from the land mixes with salt water. Water temperatures may rise and fall dramatically from season to season or even within a twenty-four-hour period. Salinity fluctuates more or less regularly, twice daily, as tides move back and forth through the river mouth, and it may also fluctuate with shifting weather conditions such as rain or drought. An animal that is subjected to water that has a lower concentration of salt than that in its body fluids will bloat as water travels from lower to higher osmolarity in the tissues. As one moves farther upstream in the brackish waters of an estuary, the water becomes fresher and different zones and population species develop based on varying degrees of salinity.

Despite the constant stress of variable salinity, life flourishes in estuaries. Cordgrasses make up the "scaffolding" for the mudflats, and beds of this high grass (also called eelgrass) are the first part of the slow transformation of shallow open brackish water into wet meadowland and, eventually, into dry land. At high tide, predators from offshore such as squid and fishes hunt between the subsurface cordgrass stalks. Many of our principle food items such as oysters, crabs, clams, scallops, and shrimp are supported by the estuary. Not only do estuaries serve as nurseries for the sandbar shark and other sharks, they provide flounder, herring, striped bass, fluke, and tarpon with the same opportunity. *Secrets of the Seas,* by Reader's Digest, stated, "Salt marshes are among the richest wildlife communities on Earth." As alluded to by Thomas Allen in *The Shark Almanac,* these precious waterways are too often regarded as wastelands or convenient natural sewers. Over the past sixty years, the dumping of dredge spoils and sewage sludge just southeast of the New York metropolitan area has created a twenty-square-mile dead zone, where worms cannot survive even on the bottom.

It seems fairly certain that at least a sandbar shark explosion occurred during the summer of 1916 in the bays of New York and New Jersey. Most likely, there were other large shark species being caught frighteningly close to shore that season as well. Are these two situations connected? Were they all attracted to the region by the same stimulus, or were the other large sharks, including the 1916 killer(s), attracted by the more indigenous sandbar sharks and their pups?

On the topic of the 1916 shark explosion and the attacks, Dr. John Nichols actually offered some reasonable thoughts. In October 1916, Nichols wrote:

> Whether sharks in general are more numerous in our waters this summer than during previous years may be seriously questioned, notwithstanding the way in which local fishermen and the crowd of incoming steamers have vied in frightening the public. Shark stories with a certain foundation and truth will always be forthcoming when reporters have been ordered to get them. It may be recalled that the summer of 1915, although marked by no such horrifying events as we have known this year, was nevertheless popularly considered an exceptional "shark season."

Of all the conjecture as to why large sharks might have gathered in a given area, the number one theory should probably be food. Above all, a shark's ability to consume adequate and appropriate prey is most vital to its existence. Most sharks are constantly moving to bathe their gills with oxygen, and they have pretty large bodies to support nutritionally. The confounding, and disappointing, fact about large sharks that are placed in aquarium tanks is that they are the most finicky of eaters. Many times the large shark will refuse to eat completely and will become progressively weak and die. The food items that aquarium curators find are most appealing to picky large sharks are fresh fish livers and, believe it or not, baby sharks. Edwin Thorne's study of sandbar sharks revealed that pregnant females gave forth their young during a stretch from June 22 to July 18. The shark nurseries of the Jersey Shore in 1916 are, therefore, to be considered a fairly tasty lure to the hungry large shark. The large shark that Paul Tarnow captured in his Keyport pound net was dissected and found to have up to fourteen baby sharks in its stomach.

Even if a large bull shark, tiger shark, or white did make its way toward the gleaming beaches of the Jersey Shore in 1916, from a strictly behavioral point of view, is it reasonable to accuse one individual of all the attacks and at the same time believe that the shark had actually developed a predisposition to do so? What we are referring to here is validation for the rogue-shark theory. In October 1916, John Nichols and Robert Murphy had to swallow a bit of pride as they wrote again of the sharks of Long Island waters for the *Brooklyn Museum Quarterly*. They wrote:

> The truly exceptional nature of the accidents should not be lost sight of. The New Jersey accidents bring the whole shark question before us in a new phase. Here in the waters for a century considered safe, we are confronted with a situation, which, in addition to actual destruction of human life, has terrorized countless numbers of people who enjoy ocean bathing, caused the loss of perhaps tens of thousands of dollars to proprietors of beach establishments, and has indeed been considered of such gravity as to be discussed in session by the Cabinet of the nation's President.

The newspapers, as early as July 8, were promoting the "solitary shark theory" because of the unique nature of the events. Such exceptional happenings were easily explained by an exceptional shark.

Interestingly, the 1916 attacks might have even seen the first real proposal of the rogue-shark theory. Just after the second attack, newspaper articles headlined MOTORBOAT PATROLS MANNED BY CREWS WITH RIFLES AND HARPOON GUNS HUNT FOR MURDEROUS SHARK [singular stressed].* The theory was not proposed after the attacks were over, and it was not proposed after the white shark was caught in Raritan Bay. It was offered not by name, but in concept, after the second attack. One day after the Bruder attack, Dr. Lewis Radcliffe, an authority on the natural history of sharks, stated that a relatively harmless shark may acquire a taste for human flesh and develop into a voracious monster, lusting for human meat. Radcliffe pointed out that there had not been an authentic record of any attacks prior to the New Jersey deaths, but he knew of evidence from Australia where certain types of sharks seemed to turn vicious and infest rivers and feed on mullet. When the mullet were depleted, they would turn on man. Dr. Radcliffe asserted that "although the blue shark or the white shark have been implicated in the New Jersey attacks," he believed it was the tiger shark. Another very accomplished resource on the subject of sharks in 1916 was William T. Hornaday, a longtime director of the New York Zoolgical Park. In his book *American Natural History,* Hornaday asserts that, "only few species of sharks are rightly classified as man-eaters. Those qualify under this classification, which are large, voracious, fearless and so aggressive that they will attack a swimmer at the surface of the water and devour him regardless of his resistance." Hornaday viewed the white shark and the "great tiger shark" as the only man-eaters.

When Victor Coppleson first offered the rogue-shark theory decades ago, shark enthusiasts either accepted it as a fantastically keen and gruesome observation or silently questioned it. Even if a researcher thought that such a theory was ridiculous, there was no adequate evidence to disprove it. The *Jaws* phenomenon would immortalize the very concept but would do little toward encouraging a scientific con-

---

Asbury Park Press.

sensus in favor of such a theory. In many scientific and popular publications, the entire theory has been avoided or discounted. According to the raw concept, certain lone sharks will, for reasons not fully understood, become predisposed to attacking man and take up and maintain a regular beat along a limited stretch of shore. Most scientists today shrug their shoulders at acceptance of the rogue-shark concept, as it is strictly presented, or simply refute it. The idea of a shark adding man to its dining list or making man the delicacy on the menu is a bit difficult to accept. I am no proponent of the "man-hunting shark" but other accepted features of fish dietary habits do correspond to the rogue view. I see vague parallels to the striped bass, for example, when variable eating habits are examined. The striped bass, which is likely the most studied fish on the planet and the most sought-after game fish by Jersey Shore beach casters, has a very distinct and frustrating pattern of eating. One day it will prefer clams, another day sandworms, the next week live herring or moss bunker, and the other times it will eat only soft-shell crabs. The striper apparently "turns on" a food choice among its "schoolies," depending on the weather conditions and/or the season and the current availability of food items. When the water is turbulent, for example, soft-shell crabs are exposed because sand is pushed off their vulnerable shells. It could be within the realm of possibility that a shark may adopt a transient new predilection for the ambulatory marine mammal called man.

A rogue sequence is defined by a sudden sequence of attacks on man in an area that was previously considered safe. The author of *Danger—Shark!* states that "such sequences of attacks have occurred too often to be matters of simple coincidence." Attack sequences that are suspicious and notable include the following: In 1922, at Coogee, near Sydney, within a two-mile stretch of coast, two attacks occurred within a month of each other and two more within the next three years. Then attacks in that vicinity ceased abruptly. In 1928, Bondi, also near Sydney, saw four attacks over a one-year period. No attacks had occurred before that time and none afterward. In the early months of 1934, an attack occurred near the George's River, and by December 31, in that same year, two people were attacked at locations three miles apart within a period of four hours.

By 1940, Dr. Coppleson had studied and analyzed enough of these gory sequences to do what shark-attack researches have always endeavored to do: predict shark behavior. The prophesy came about after a dog and a man had been attacked at Brighten Beach in Botany Bay on January 23. Coppleson wrote a letter to the *Sydney Morning Herald* and warned that if the shark responsible for the unusual attack events was not caught, another attack in the same area should be expected. Twelve days later, a second man lost his life only four hundred yards from the first attack. Close to forty years ago, in *Danger—Shark!*, Jean Campbell Butler wrote: "Sometimes the rogue shark appears not to patrol a single limited beat but to cruise along a more extended path, killing and maiming as he goes." That statement was Butler's introduction to a short discussion of the classic example of what was termed the "long-range cruising rogue." Butler was referring to the Jersey man-eater.

Should the rogue-shark theory be discounted? Is it merely a fantastic form of conjecture and speculation that sensationalizes and fabricates a connection between unrelated freak attacks? Perhaps the attacks occur in a given location or in a given time period because of occult environmental factors (e.g., illegal dumping, offal disposal), food availability alterations, or unappreciated features that attract dangerous sharks. In reference to unappreciated factors, the summer of 1914, in contrast to 1916, was the coolest summer of the entire century. How might that have impacted on subsequent summer seasons and sharks?

In 1916, a member of the U.S. House of Representatives even got into the act of proposing theories on the man-eaters. Congressman Peter F. Tague, of Massachusetts, theorized that steamship companies, restricted by the wartime situation, had been dumping horse and cattle carcasses nearer to shore. Tague challenged the secretary of the Treasury and the secretary of Agriculture to direct all steamship lines to restrict dumping to locations far off the mainland. The basis of Tague's theory would certainly have a bearing on shark activities, if his accusation was accurate.

Regardless of whether the entire 1916 attack quintet was the work of one shark, it would be quite surprising to find that the damage done at the Matawan Creek was not the result of a solitary shark. Indeed, that one hour of toothy creek terror could stand as the most incredible short-term rogue sequence the world has ever known. It is not surprising, therefore, that even the most modern and state-of-the-art, shark-schooled specialists put at least some credence into the feasibility of rogue series. Geremy Cliff, for example, current director of the Natal Sharks Board, believes that it would not be surprising to find one large shark to be responsible for an abrupt string of attacks off beaches that were previously undisturbed or where such a slew of attacks suddenly starts or stops.

After a shocking attack on August 30, 2000, in the Gulf of Mexico, near Tampa, Florida, I took the opportunity to contact George Burgess about his feelings on the rogue-shark theory. Burgess is considered the United States's foremost authority on shark attacks and is currently the director of the ISAF and the senior biologist in icthyology at the Florida Museum of Natural History. The attack near Tampa saw a sixty-five-year-old man lose his life while swimming near a dock in five feet of water. A large shark came at the man in full view of his wife and mauled him in the armpit, the leg, and the chest. He died immediately, and Burgess used the witness descriptions and the characteristics of the wounds to implicate the bull shark. The tiger shark is known to inhabit the northern Gulf during the summer, but Burgess did not feel that the tiger shark was a probable culprit. The shark was reportedly eight to ten feet in length. The Gulf side of Florida is an area that sees fewer attacks than the east coast of the state, but when attacks occur, they are usually perpetrated by larger sharks and are usually more serious. Fatal attacks in Florida are exceedingly rare in general.

The fatal attack by a large shark on the Gulf side of Florida was a perfect prelude to our discussion on the rogue topic. In regard to the ISAF and the 1916 attacks, Burgess simply stated that his review of the documentation and the analysis done by his predecessors at the ISAF led to the conclusion that a white shark perpetrated the acts in 1916. Burgess set the tone firmly, however, on the subject of rogue sharks. He conveyed to me that, from a strict behavioral point of view, the rogue premise was generally not accepted by authorities in the field, including him. The idea that Coppleson attached a "man-hunting" shark to the sometimes prolonged (e.g., years) sequence of attacks in designated rogue attacks was just unacceptable. When the topic turned to the question of the 1916 New Jersey incidents, Mr. Burgess showed a respect for the uniqueness of the string of Jersey maulings. "In the 1916 New Jersey case," he said, "you have a sequence of attacks which occurred in a south to north fashion in a reasonable period of time, with a white shark captured in a northern region of the most recent attacks, with human remains in its stomach, and a stop to the attacks after the capture." George Burgess would not go along with calling the New Jersey culprit a rogue shark, but he did say, "in this one particular case it is not unreasonable to consider one shark as the responsible attacker."

The consensus among shark authorities is that rogue sharks do not exist. I suspect it is the element of turning a shark into a "Dracula of the deep" that turns off the modern expert from accepting such a behavioral and philosophical leap. On this very subject, it is quite intriguing to note that two of the three authors of April 1916 Brooklyn Museum paper, Robert Murphy and John Nichols, collaborated on an October revision of their earlier article and wrote with an uncanny modern logic in their conclusions about the 1916 attacks. While these men said that the Schleisser white shark was probably the killer in all the 1916 attacks (while Dr. Frederic Lucas felt that the attacks were related to a "shark year"), they could not have believed in a rogue concept when they wrote:

No shark lives upon human flesh or even has an opportunity to "acquire a taste" for it. Even the great white or tiger shark doubtless takes what lowly food he can get with the least exertion. . . . Human bathers naturally offer a most abundant and accessible food supply along our beaches, and the peregrinating man-eater merely exercises his ancient privilege. . . . The chances of being attacked by a shark are still infinitely less than that of being struck by lightning, though this is, to be sure, small comfort to the victims.

In a sense, the New Jersey attacks may be the only true *roguelike* (I will refrain from calling it a rogue here in respect for scientific accuracy) attack sequence in history and just happens to be a long-range cruising version. The simple fact is that some sharks might present as behaviorally aberrant to the point of sequential attacks. We have long known other marine species like blackfish whales to ground themselves to death and later find that their gastrointestinal tract is filled with parasites. On land, bears and predatory felines that turn inordinately vicious are often found to have some type of simple ailment that explains their anomalous behavior. From the standpoint of heredity (which plays a role in the accessory rogue discussion forthcoming), as opposed to strict experience and conditioning, the runt of a litter may also be the most docile or the most aggressive because of a genetic or pragmatically encouraged predisposition. This is why the general dictionary meaning of a rogue can include "a chance biological variation." When I confronted Burgess with the fact that large dangerous sharks are commonly present near human bathing regions without going on "five attack rampages," he resorted to a clever human analogy: "Many people walk around with guns and don't harm anyone, yet once in a while one of them ends up on the top of a building and starts shooting at people."

When a prospective rogue sequence has a reasonable temporal association (i.e., over several months as opposed to over several years), I consider the rogue, or at least identifying a solitary shark in multiple attacks, within the realm of scientific/animal behavioral possibility. When attacks are artificially strung together (blaming the same shark) over a period of years, however, the cohesion of the attacks and the solitary nature of the perpetrator becomes questionable. Other authors and researchers may still consider it quite possible that an individual shark, even after years, after having consumed a piece of man or at least aggressively "mouthing" a human with some satisfying or innocuous result, will be more inclined to "reapproach" man as a potential prey item. Such encounters may occur over years, months, or *twelve days*. We all know the nasty dog in the neighborhood that has snipped at a person or two and gotten away with it. Sharks are obviously not as intelligent as dogs, but they are capable of learning. Learning is, in its broadest sense, an adaptive change in behavior that results from experience. A baby toad, for example, may strike at a bug with its tongue as a result of a stereotyped prey-capture behavior. Experience comes in when the toad must alter its feeding habits according to environmental features (e.g., avoiding poisonous insects or striking at new tasty/prevalent ones). The change endures for some time and is adaptive (it makes biological sense for the toad to eat certain prey items). The shark, like the toad, may trace its altered behavior to a discrete event in its lifetime.

Among other species of the animal kingdom, the rogue phenomena or rogue pattern does seem to occur. The murderous rogue tigers, lions, or elephants that somehow become socially and behaviorally estranged from their own species and normal practices have been known to terrorize certain villages in India or in Africa. In these cases, it is considered more an aberration in the individual creature rather than any factor in the environment which converts or forces that particular specimen to act out in such a fashion. It could be actions that are secondary to an impairment or a physical or mental disease (even rabidlike) which make the shark incapable of pursuing its natural prey. The human correlate (if one exists) might be akin to a physically or psychologically deprived person who must resort to illegal activities to survive, or in the extreme case, the freakishly deranged person who perpetrates mass murders.

I cited the strange case of Raymond Short in Australia, in which a young white shark affixed itself to his leg until the shark was forcefully released on dry land. This attack was compared to the attack of Charles Vansant. In fact, there are at least three instances in the ISAF, besides the Vansant account, that refer to a shark affixing itself to a victim's extremity. In the Raymond Short case, the white shark was found to have massive wounds of fairly recent origin. The shark's left and right lower abdominal surfaces and its dorsal and ventral regions of the tail and pelvic and caudal fins were effected. One wound on the shark was clearly the result of a bite from a shark of equal size. The abdominal cavity of the shark had been penetrated by one large triangular wound, which was large enough to fit a finger. Despite the wounds, the shark still had food remnants in its stomach. Could the shark that attacked Raymond Short have been injured to the extent that it could no longer capture its natural prey?

It was Victor Coppleson, a fellow physician and shark-attack analyst, who was the sole early proponent of the entire rogue premise. Another shark expert, however, also sought to explain the behavior of atypical individuals that were apparently divorced from the normal life of their group, erratic in feeding habits, prone to lurking in shallows, and especially dangerous to man. The parallel theory was that of accessory populations and was proposed by the most prominent shark biologist of the past century, Stewart Springer. In *Discovering Sharks,* George Burgess wrote:

> During this century, the field of elasmobranch studies has been blessed by a number of prominent workers—the names Garman, Bigelow, Schroeder, and Gilbert immediately come to mind—but one person stands out above all others when discussions turn to sharks. . . . It is safe to say that there isn't a serious student of sharks who hasn't heard of Stewart Springer.

Listing Springer's credentials would be an exercise in reciting superlatives, and the reader should safely assume that his educational background and his professional and occupational positions in the field of shark biology might be outweighed only by his enthusiasm and his onsite observations.

As far back as the early 1940s, Springer realized that the exceptions to generalization in the sphere of shark behavior were certainly frustrating to the analytical observer. In an attempt to explain contradictions or exceptional migratory movements or life patterns of certain species, he developed the theory of principal and accessory shark populations. Fortunately, for purposes relevant to this analysis, he looked at the total population of migratory large sharks in the temperate and subtropical western North Atlantic region. The principal population was defined as the core group and the main breeding population whose habits and existence maintained the abundance of the species and followed regular patterns of distribution.

The members of the accessory population, on the other hand, are our recruits for the small and select world of rogue sharks. These individuals are said to be disoriented, cut off from group life, displaced, and/or usually permanently lost. They may have strayed, by chance, out of the normal geographical range of the species and may be cut off from the natural phases of group life, including seasonal movements and cyclical reproduction. The numbers in this accessory group fluctuate but are assumed to be low. Springer also believes that evidence supported the conclusion that the percentage of attackers (against man) among sharks of accessory populations far exceeded the proportion that their number should contribute. He did not assert that sharks of principal populations were harmless, but that their "normal" behavior made them predictable for that species. He was, in essence, saying that an accessory individual was, by nature, unpredictable.

Springer's observations and analysis of shark-fishing reports demonstrated a few keys in identifying these accessory sharks. Most important, these sharks were said to be the ones that were present when the bulk of their species was known to be in other locales. Such individuals had the following tendencies: they would be the sharks most apt to be caught near harbor mouths and in shallow waters; they preferred the company of *boats;* they were easiest to catch even with spoiled bait; and they were either very young or very old. More members of the accessory population showed major injury or deformity than did the principal population. Where the accessory theory completely departs from the rogue human-predilection concept is where both populations (accessory and principal) seem to have the same food preferences and aversions.

Inserting certain shark species into the accessory equation does become somewhat problematic. The tiger shark, for example, is often distributed in solitary arrangements and is naturally and commonly found close to docks (boats), refuse, and in other unusual inshore locations, so accessory population distinctions are almost impossible. Because of the white shark's limited numbers, well-defined patterns for their species are very difficult to obtain from direct observation. The bull shark variant of Lake Nicaragua, however, may be an accessory population in itself. This variant is easily caught, is known to attack man in shallow waters, is indiscriminate, and is bold and voracious.

In the last minutes of my conversation with George Burgess, I couldn't wait to ask him what became of the groundwork that Stewart Springer laid in proposing the theory of accessory populations, and what scientists think of such a theory today. To my surprise, the answer did not point to dismissal, confirmation, or expansion. The short truth is, because of technological limitations at the time of its proposal, no one followed up on it. Burgess was optimistic, however, that advances like mitochondrial DNA mapping will identify the truth about subpopulations and follow through on some other theories proposed by Springer.

It was a good idea to "check with Stewart Springer" before concluding this investigation, and I believe that Springer's observations and theories may well be the tree that will bear fruit on the branches of the new millennium's technological prowess.

Beyond the theory of rogues or accessory populations, noting the white shark's presence as an occasional inshore predator does have relevance to our discussion on what transpired in New Jersey in 1916. Tim Wallett, author of *Shark Attack and Treatment of Victims in South African Waters,* theorized that sharks inhabiting inshore waters have evolved with an abundance of food. This food is always at hand, and the sharks normally have not developed excessively aggressive behavior. In the pelagic environment, however, food cannot be located as easily. Wallett states that the prey items in the pelagic sphere "must be hunted, attacked and then consumed. . . . To survive, these sharks have had to evolve a savage disposition, seizing the chance to feed whenever the opportunity presents itself." The great white happens to be the only pelagic shark (with the mako as a very rare exception) to venture into inshore gillnets off Natal, South Africa. The white shark is also known to enter the bays of New Jersey and Massachusetts in the United States. Wallett additionally stated that the white, with its large teeth and aggressive nature, is responsible for many of Natal's attacks (as opposed to those who see the bull shark or general carcharhinids as the overwhelming culprits). In his 1978 publication, Wallett wrote of the white, "As long as it remains in nearshore waters it can attack bathers."

In light of the behavior and knowledge of the "inshore white" I wrote, in 1987, that though the New Jersey summer waters may contain a number of dangerous adult whites and carcharhinid sharks, the majority of its past attacks may have been the work of large *juvenile* whites, with dietary habits in a transitional phase (from demersal fish satiety to pursuit and desire for marine mammals). Here, these "transitional white sharks" had accidentally strayed inshore during heavy human aquatic activity.

As Nichols and Murphy stated in October 1916, the simple presence of "harmless" sharks in summer waters off New Jersey and New York was something people should get used to. They intimated that, from time to time, a treacherously "venomous serpent" (man-eater) may be among them:

> For the citizens of the Northern Atlantic states, our sharks are in a very real sense analogous to our snakes; of each group you have many common inoffensive kinds, and exceedingly few, rare, dangerous kinds. We should learn to take for granted the presence every summer of ground sharks, sand sharks, and threshers, just as most of us accept without concern the harmless garter snakes and puffadders of our woods and fields.

With all due respect to these authors, even the harmless species can cause a problem when threatened; I distinctly recall as a child hunting for frogs at Deal Golf Course when we happened upon a garter snake that gave us a very threatening open-mouthed (toothless, I might add) hiss and display of aggression.

Women ready for "fanny dunking." Early 1900s.

CHAPTER 10

Just Beneath the Surface

In the two decades after the 1916 attacks, the stir surrounding the potential danger of sharks subsided. Attacks occurred, but the incidents were more easily attributed to euphemistic causes such as a stray "large fish." It was scientifically and economically safer to dismiss the infrequent attack as a unique nonevent rather than create negative publicity or controversy. The 1916 attacks became conveniently stored in archival newspaper files and the dormant memories of all who were living at the time of the Jersey man-eater's reign. The only flash of reminder to come to the New Jersey beaches in 1917 was a minor shark-inflicted gash sustained on the knee by lifeguard Daniel Thompson at Seabright. That incident occurred after Labor Day, September 21, which was way too late in the season to get anyone alarmed. The United States was then involved in the war, and the sharks would take a backseat for decades.

Twenty years passed, and it was time for the old Mount Vesuvius of the sea to gurgle once again. On July 25, 1936, near Mattapoisett, 150 yards off Hollywood Beach, on Buzzards Bay, Massachusetts, an undeniably shark-inflicted death occurred. Sixteen-year-old Joseph C. Troy

Jr. was doing a freestyle crawl while a friend, Walter Stiles, six feet away, was doing a quiet sidestroke. Suddenly Stiles saw Troy being pulled under by his left leg. Stiles screamed to a boat closeby, and since Troy was too weak to grab hold of Stiles, Stiles placed him in a rescue hold and swam him toward the approaching rowboat. In the boat it was revealed that his left thigh, from hip to knee, was cut to ribbons. Joseph also had lacerations on both of his hands (from hitting the shark), and he had lost a finger. As the boy was hauled into the boat, it was noted that a large shark came within three feet of the small vessel, swerved, and went under. The boy was taken to New Bedford Hospital, eleven miles away, and during efforts to amputate his leg, he expired from blood loss and shock.

Dr. Hugh M. Smith, formerly of the U.S. Fisheries Bureau and one of the original skeptics of shark involvement in the first 1916 attack, was summering at Woods Hole on Buzzards Bay in 1936. Smith reported that, after reviewing the comments on the shark's coloration and tail shape, he believed the perpetrator at Buzzards Bay to be a great white shark. Smith stated that the white shark had been spotted in the bay some twenty times by a man named Vinal Edwards between 1871 and 1927. It was E. W. Gudger, however, of the American Museum of Natural History, who generated a formal analysis of this attack and related its style to that which was seen during the New Jersey attacks. Gudger was quite right. The attacks on Vansant and Fisher, twenty years before, were very similar to the one that took the life of Joseph Troy.

Although Gudger's and Smith's statements on the Buzzards Bay attack were without restraint as to the shark's responsibility, the men did not seem to feel that a second attack would be likely. Whenever a serious attack occurs anywhere in the world, it is relatively safe (and helpful to local commerce) for the specialist to downplay the chances of a recurrence. Gudger himself likened the chances of being attacked in that particular area equal, once again, to being struck by a bolt of lightning. Indeed, after the 1936 attack, decades would pass before attention to shark attacks would materialize with any force. The horrifying tale of the cruiser *Indianapolis* in World War II and the gruesome attacks in South Africa in the late 1950s would finally turn the tide once more.

In reference to attacks in the northeast United States, however, Gudger's statement about the exceptional nature and exceedingly rare occasion of shark attacks in that region would prove correct. Even though tasty harbor seals frequent Buzzards Bay today, along with not-uncommon sightings of the white shark, no one has been attacked since Joseph Troy. Richard Ellis points out in *Great White Shark* that "since the 1936 incident, there has not been a documented white shark attack in northeastern waters, even though great white sharks have clearly been present." Although not in the United States, but still in North Atlantic waters, a lobster dory was viciously attacked and sunk by a white shark off Cape Breton Island, New Brunswick, Canada, in 1953. The incident was immortalized in a painting by Paul Calle.

As I've mentioned earlier, almost every life-threatening attack is committed by a shark that is near or greater in size to a human being. The attacking shark is also invariably dark in appearance. Beyond those general observations, we get little more that is accurate or specific from the understandably unreliable witnesses. The excitement and rapidity of the attacks and the overt concern for the life of the human victim always distorts the clarity of the observation. Therefore, to delineate discrete tail features or attack patterns to the liking of the hungry shark-attack authority is virtually impossible. Elucidating material like tooth fragments, wound characteristics, and location are tremendously helpful, but the tooth fragment is the only one that will determine the species beyond doubt and, unfortunately, the telltale tooth is rarely found.

On the stretches of U.S. East Coast north of Florida, attacks have indeed occurred since 1916. In no way, and in no part of this book, would I want to give the reader the impression that I am maligning the magnificent endangered white shark by trying to implicate it in attacks of which it has been innocent, but I do want to convey the reality of its natural behavior and the fact that, just like the polar bear, it must be respected for its life-threatening facilities. For reasons such as the limitations in identifying an attacker, the decline in active reporting to the ISAF in the 1970s, and the economically disruptive publicity that can follow a white shark attack, I would not be surprised if the culprit in some, if not many, of the Northeast's attacks over the past fifty years was the white shark.

There have been forty-one reported attacks along the southern East Coast of the United States, above Florida, in the past century. The bulk of these attacks, thirty-one, have occurred in South Carolina, and the sharks most often blamed are the blacktip and spinner sharks. These attacks have not normally been life-threatening. Above North Carolina, from Virginia to Massachusetts, fifty-three attacks have been reported since 1916. From the total for the century, Connecticut reports two, Delaware four, Virginia five, Massachusetts seven, New York twelve, and, as you could guess, the infamous distinction of having the most goes to New Jersey, with twenty-eight. It should be noted that the spinner shark and the blacktip shark do not venture above North Carolina.

Although the number of attacks in the Northeast, and especially in New Jersey, is substantial, it pales in comparison to the number of attacks in Florida, California, and other regions of the world. Additionally, while there has not been a *documented* attack in New Jersey since the 1960s, the other areas of the United States prone to attacks, as well as some other regions of the world, are actually seeing fairly dramatic increases in attack numbers. The white shark attacks in California are pretty clearly attributable to the increased density of marine mammals (which whites feed on) since the inception of the Marine Mammal Protection Act in 1972. In Florida and the other areas of the globe, the increase has been blamed on the obvious increase in human inhabitants or commercial fishing activities (shark luring) close to shore.

The incremental rise in attacks during the period 1980 to 1999 (805 worldwide) and even over the past century is directly related to world population growth and increases in human utilization of marine waters. The apparent drop in attacks in the 1970s is said to be related to the largely inactive state of the ISAF. The all-time high in unprovoked attacks for one year was seventy-two in 1995. The United States was the clear leader in attacks with thirty-four, followed distantly by Australia with five, Brazil with four, and Bahamas and South Africa with three each. There were a scattering of attacks in other parts of the world at that time. Matthew Collahan, an expert on attack trends, said that the 1995 number reflected the availability of sharks and the amount of time people spent in the water. He speculated that warmer temperatures might be the simplest reason to explain both factors and that surfers are the most targeted victims, with nearly half of all 1997 attacks involving surfers, wind surfers, and rafters.

It is interesting to note that the northeast United States and the New Jersey coast have not been involved in this incremental increase in attack numbers, even though its waters are certainly filled with thousands of summer bathers and many year-round surfers. I do have to admit that I have received anecdotal accounts of minor attacks during the 1970s that were not reported. One example was an attack that occurred off a dock in the Navesink River in Shrewsbury, New Jersey. A man was dangling his foot in the water. When he pulled his foot from the river he found he was missing half of it. Beyond such incidents and the attacks of 1960 (which we will examine in detail), New Jersey has not seen an increase in attacks. It should be pointed out, however, that unlike Florida and the southern United States and the somewhat static water temperatures of California (not to mention California's exploding pinniped population), New Jersey has only two months (July and August) in the year that sees the combination of warm (seventy degrees Fahrenheit) seas and dense human water recreation. In contradistinction to the other attack-prone areas in the United States and the world as a whole, New Jersey has a shorter stretch of coastline and many more cold-water months (waters below fifty-five degrees Fahrenheit, with March water dipping to the upper thirties). It also has a paucity of nearshore sharks over the cold period. Early March in New Jersey probably represents the coldest sea temperatures. Since it takes a great deal of time and energy to exert a change in water temperature, the early March period is the time when winter has driven the sea temperatures to an annual low. My brothers and I are avid New Jersey shipwreck divers, and we often get brief but vicious "ice-cream headaches" when we venture below the thirty-foot mark in late February. I vividly recall dissecting a sixty-foot fin whale that had floated ashore in Deal, New Jersey, in March 1986. The dead whale had been floating at sea for several days, and when we examined it, we found not one shark bite.

The decline in shark attacks in New Jersey may also relate to some dramatic declines in the general shark population. After some northeast attacks in the 1960s, the U.S. government and state governments began encouraging shark fishermen to revive shark fishing as a way to reduce damage by sharks on commercial fishing gear. Additionally, in the 1980s, government nutritionists urged the catching of sharks to bring an "underutilized" source of protein to the American diet. The U.S. commercial fishing of shark came to an all-time high in the 1980s and early 1990s when, still again, the NMFS encouraged the procurement of shark fins for the foreign shark fin soup market. This was suggested to relieve the pressures on other declining U.S. fish species. For decades New York and New Jersey have held annual shark-fishing tournaments. These tournaments alone could affect shark population and distribution.

American Atlantic waters were recognized as the last great reserves for sharks. In 1997, however, the NMFS declared that the Atlantic shark populations were in a precarious state and did all they could to chop down the catch limits for large coastal sharks. Of course the most endangered of all shark species is the great white. As Doug Perrine said so fittingly in *Sharks and Rays of the World,* "An ocean without sharks and rays will be like the plains of Africa without lions and tigers, like the mountains of North America without wolves, or like the Arctic without polar bears . . . reducing the magnificent tapestry of life by degrees to a ragged shawl."

Despite the decline in northeast attacks over the past thirty years, 1960 revived memories of 1916 from the dusty stacks of microfilm. It all began in late August of that year.

Each Jersey Shore summer comes to a slow halt with the arrival of tropical water and magnificent waves created by the passing August/September Atlantic hurricanes. During mid-August of 1960, Hurricane Cleo had just passed by. There is nothing closer to paradise than the west wind setting that comes after a low pressure system moves offshore. Such a combination of events creates a sun-filled sky and glistening mountainous surf.

On August 21, 1960, a twenty-four-year-old man had just made his way down from Jersey City to Sea Girt, the town immediately south of Spring Lake, to reap the rewards of the spectacularly refreshing surf. His name was John Brodeur. In 1959, a Spring Lake woman was said to have been pulled from the path of a wounded shark by two lifeguards. On this day, however, the waves were great for riding, and sharks were the furthest thing from people's minds. At 3:30 P.M., John's fiancée, Jean Filnamo, was wading closer toward the beach while John made his way into the crowded deeper water. John entered the section that was used by the hotel employees from the nearby Stockton Hotel. Four decades before, Charles Bruder swam in the employee section as well.

John, a young accountant, and Jean, a Jersey City schoolteacher, were set to be married September 17, and the invitations had just gone out. They waited for the hurricane to pass the coast before coming down to the shore, but the water was still churned and murky. Fishing-party boats out of Belmar could be seen chumming just off the beach. The boat captains were unable to negotiate the still-treacherous larger swells offshore. Brodeur was about seventy-five feet from shore and had taken several great rides. He was a proficient bodysurfer and found that the location he identified, which was waist- to chest-deep, seemed most reliable for that tide. From that spot, he could almost predict the big riders. As one wave went by, Brodeur spotted a blackish object sticking out of the water, about ten yards away. Anyone who is experienced with riding in poststorm surf knows that dock timbers and other varieties of driftwood can end up near the beach in the subsequent days. Brodeur was not sure if the object was a log or a body surfer taking a wave toward the shoreline. It turned out to be neither.

Just seconds after seeing the "blackish object" disappear behind him, Brodeur planted his feet to exert a push to get into an oncoming wave. At that very moment, he said, "I was hit with such a force that is indescribable. I could possibly liken the crushing blow from behind as from a Mack truck and then a jerking pull onto my right leg. I used my right leg to feel at something rough below and I began to hit it with my hand. I felt no immediate pain," he recalled, "but the water around me turned an instant red." He yelled to his fiancée, and she and three men, including his future father-in-law, came to his aid. On the beach, he and the onlookers were immediately horrified at the obvious gravity of the situation. A first-aid squad speedily made its way toward the beach and arranged for a clear road path to race Brodeur to Fitkin Hospital (now Jersey Shore Medical Center). The ambulance driver, Brodeur remembers, used quite a few expletives to describe his frustration when they eerily found the Shark River drawbridge elevated to allow passing boats to come through the channel.

At the hospital, Dr. Charles Samaha, a general surgeon, did all he could in a four-and-a-half hour operation to revascularize and surgically preserve what was left of Brodeur's lower leg. Samaha, a family friend, informed me that the wound was enormous and crescent-shaped. The shark's teeth had penetrated through the fibula and halfway through the tibia. With Labor Day weekend approaching, the local economy was in a tenuous situation regarding publicity. The onslaught of local reporters and resort officials even tried to encourage Dr. Samaha to consider blaming the incident on a stray blimp propeller. As you remember, the Jersey Shore mayors in 1916 tried similar stunts.

Brodeur's father, Edward Brodeur, rushed down from Jersey City to be at his son's side. Mr. Brodeur was originally told that both of his son's thighs and calves had been mutilated. When he arrived, he was relieved to find that it was only one calf. Despite the finest medical care of that era, Brodeur was so weak that he could not even carry on a conversation after the surgery. Within days, the stench in Brodeur's hospital room, his toxic appearance, and his spiking fever told the tale of a lower leg that was progressively becoming gangrenous and infected. With his fiancée at his side for photos, Brodeur and the physicians had no choice but to perform a below-knee amputation. Brodeur recalls that immediately after the putrefying portion of his lower extremity was removed, he felt "one hundred percent improved."

During the filming of *Tracking the Jersey Man-Eater*, Dr. Robert Patterson mentioned a bizarre detail relating to the Brodeur attack of 1960. The jovial doctor told me that Brodeur was actually faced with a dilemma after the surgical amputation. The question was whether to bury his lower leg, with a Christian ceremony or dispose of it through the usual hospital standards.

When Brodeur came to my home in Allenhurst in the summer of 1998, he was joined by his pleasant second wife of many years; he was happy to tell me that he still gets in the water to bodysurf and wasn't going to let one incident keep him from enjoying the marvelous Jersey sea. He also informed me that he continued to correspond in a warm fashion with the surgeon, Charlie Samaha. Before Brodeur departed, I had to ask him about the leg-burial story. "Yes," he stated, "as a practicing Catholic, I decided to have the leg buried in its own coffin. Whenever I pass the cemetery near the hospital," he said with a smile, "I tell my family that part of me is buried there."

The ferocity of the Brodeur attack may have inspired some to question the potential for another attack. One day after the Sea Girt incident, at Seaside Park, ten miles to the south of Sea Girt, fourteen-year-old Thomas McDonald reported that he was bitten by a shark. For medical reasons or public relations reasons, the boy's shark story was not given full credibility. He had an injury, but according to the newspapers and the emergency room physician, it was not caused by a shark (sound familiar?). The boy's laceration was not serious, and the event was not publicized. On August 24, at a different Seaside Park, near Bridgeport, Connecticut, Clyde Trudeau, a thirty-eight-year-old snorkeler, was seventy-five yards offshore when he sighted a fin in the water. Within seconds he was attacked. Trudeau received a jagged laceration on his left arm, but no particular shark was identified.

The shark-alert panic was beginning to come to a slow boil on the Jersey coast and in the Northeast. Most people were just hoping that the summer would end before any "strange accident" could occur. On August 29, eight days after the Brodeur attack, a twenty-five-year-old medical student was tempting fate by swimming nearly a quarter mile offshore at Ocean City, New Jersey, fifty miles south of Seaside Park. Richard Chung, like Charles Bruder forty-four years earlier, screamed loudly and caught the attention of the lifeguards on duty. Richard Clune and Lawrence Stesdem went to Chung's aid in the lifeguard boat and, like Bruder, Chung was heaved into the boat screaming, "I've been bitten! I've been bitten by a shark!" His leg was bitten to the bone, but the physicians at Shore Memorial Hospital were able to save the limb. The type of shark that attacked him is unknown.

Now, at the tail end of August 1960, within a stretch of eight days, two or perhaps three New Jersey swimmers and one Connecticut diver were victims of shark attack. The scare was great enough to trigger armed helicopter patrols and Coast Guard lookouts. Lakehurst Naval Air Station, the same facility that saw the *Hindenburg* dirigible disaster of 1937, was now sending off blimps to help hunt for sharks. The technological advances in 1960, as compared to 1916, are obvious. Supris-

ingly, though, the reaction of the press and shark hunters was not en-
tirely unlike that of 1916. The headline of the August 25, 1960, *New
York Times* story relating to a Coney Island shark incident read: POLICE-
MEN PRACTICE HUNTING SHARKS, THEN RESCUE BOYS IN ROWBOAT. The fer-
vor and substance of that headline would have fit into the 1916 sensationalism perfectly. The only difference in this modern hunt was that
the police were flying around in helicopters and using submachine
guns. Besides carrying rifles in August 1960, the Coast Guard vessels
were dragging raw meat, just like they did forty-four years before. Jack
Casey and the National Marine Fisheries Service were enlisted to undertake a massive field study (in 1916 it was called Shark Hunt General) off New Jersey and Long Island to determine the existing shark
populations. The study garnered over three hundred sharks, with a
surprisingly large number of great whites and tiger sharks, including
an eleven-hundred-pounder (and no, I do not believe one bull shark
was taken).

The string of 1960 attacks even inspired a renewed intensity among
attack specialists to make sense of the unpredictable flurry of shark
events. Instead of stainless-steel wire-mesh fencing or semicircular piling barriers, an experimental bubble screen was installed in the
bathing area of Sea Girt. The bubble screen was essentially a long garden hose with intermittent perforations connected to a pressurized-
air supply. It was vended by Fran Arpin of Orange, New Jersey, and was
demonstrated at the Sea Circus at Asbury Park (almost reminds one of
the shark exhibits of 1916). Dr. Perry Gilbert, however, determined
that this device was only inconsistently effective. Regardless of the efforts to get scientific opinion on the 1960 attacks, the Jersey Shore officials were more concerned about the safety and mind-set of New Jersey bathers (or prospective bathers). They were concerned that the
bathers would actually make matters worse by exaggerating an alarm
about the threat of sharks.

The local press did all it could to remedy the situation. After all, no
tourists meant no big newspaper sales. Articles were even printed

about spotting dolphins off assorted locales, like Monmouth Beach and Bradley Beach, and the headlines preached how the presence of the dolphins confirmed the fact that no sharks were around.

While the 1960 attacks were viewed, understandably, as an unfortunate string of events, for our purposes these attacks are helpful in continuing to shed light on the causes of the 1916 attacks.

As already mentioned, New Jersey has not seen any consistency in attacks since the 1960s, but if its shores were to influence a higher rate of attacks compared to other mid-Atlantic states, it would probably be due to its 127 miles of heavily populated beach-lined summer coast, its strips of wide underwater sandbars, its intermittently murky surf water, its multitude of shallow water demersal (bottom) fishes, its potentially irresponsible longlining crews that may venture close to shore, and its inlets and bays sought after as shark nursery and migratory grounds.

In the quest for a culprit in the 1960 cases, the vague attacker descriptions certainly open the door to several dangerous sharks, including the white and the tiger shark, especially in view of the concurrent field-study catches. In the attack on John Brodeur, two factors besides its being described as a "blackish shark" (which whites certainly can be when wet) stand out that point to its being a white shark. The water was churned and murky that day as an effect of Hurricane Cleo, and Brodeur observed a portion of the long axis of the shark (he thought it was a log or a bodysurfer). In the foamy water, the shark very well could have raised its head out of the water, perhaps to "eye" John Brodeur. Remember, white sharks are the only sharks known to lift their heads out of the water to identify prey. Brodeur's attacker also appeared quite comfortable pursuing its prey in and among large waves, just as white sharks chase down seals and sea lions that ride and play in surf. Possibly my favorite painting by Richard Ellis illustrates a large white shark, fully out of the water, parallel and inside the curl of a ten- to fifteen-foot wave, in pursuit of a California sea lion. The sea lion is breaking the crest of the wave in the painting, and the shark eyes the frantic creature while getting tubed like a surfer.

Regardless of the frequency or scarcity of attacks off the Jersey Shore in recent years, a more important question may relate to whether increased numbers of large sharks were ever identified nearshore during any one season. Such an occurrence was documented for 1916, but as for an overt "explosion" of inshore sharks, I am unaware of that ever occurring in modern times. I am aware of stingray explosions, but not shark explosions. Although a shark population boom has not been known to occur off New Jersey since the 1916–17 period, the number and species of sharks captured during the NMFS study inspired by the 1960 attacks was somewhat surprising and worrisome to the experts. Jack Casey and his colleagues were especially stunned with a catch in 1964. The event would not be classified as a population or distribution explosion, but it is so unusual that it may be even more important to the 1916 case than the implications of the sandbar shark invasion. The event I speak of relates to the presence of a group of white sharks in a certain location off the coast of, you guessed it, New Jersey.

Jack Casey and his associates have plotted the distribution of the white shark in the mid-Atlantic bight and conclude that the white usually swims alone, mainly unassociated with other individuals of its kind. The white shark here, like in other parts of the world, appears as a loner, or at least as part of a string of sharks scattered over several square kilometers, not often assembling in schools or even joining in pairs. It's almost as if each white shark is an apex predator that prefers to hunt alone rather than reap the benefits of social systems and hierarchies. We've spoken in general about the shark's inability to withstand artificial pressures (such as being caught and killed by man), and the issue of the white shark's distribution patterns may make it the most vulnerable of all. Obviously, whites pair up to mate, but the bulk of their history is quite a puzzle. It's almost as if the white shark is prone to live the life of an "accessory shark," similar to the tiger shark. If whites and tiger sharks had a predisposition to become accessory individuals, that would present a problem for the whole species. According to the theory, the accessory individuals are not gravid.

Regardless of the white's reclusive propensity, at least one exception to that pattern has occurred. The anomaly took place off Sandy Hook, New Jersey. The finger-shaped peninsula of Sandy Hook is the white, sandy, northern beginning of the Jersey Shore. The hook arcs toward New York Harbor, and just west of its beaches are the bayside (Sandy Hook Bay) and the Raritan Bay. The hook has a rich Revolutionary War heritage, a Coast Guard history, a famous marine sciences laboratory, and the oldest working lighthouse in the United States (commissioned by General George Washington). Equal to its heritage, Sandy Hook represents a National Park and provides magnificent white sandy beaches. On many a hot summer day, the hook is so crowded with beachgoers that it must close its tollgates.

In August 1964, Jack Casey and other biologists working out of Sandy Hook's marine lab were performing shark-population studies. The men decided to drop their longline near the beaches at the beginning of their trip rather than at the end. They would not want to lure any large or potentially dangerous predators toward shore on an inbound trail, especially not off a populated beach area. The men did notice, however, that bluefishing (*Pomatomus saltatrix*) boats had just had an abundant catch not far away, and the mates on the vessels were discarding the fish heads and entrails in the vicinity of their longline route. The men decided not to take multiple passes, but rather head out to deeper waters. When they began to haul the line in they knew that fish were on, big fish. The first hooked specimen neared the boat was immediately visible as a shark. In the water, it looked like a mako, but to get a pelagic mako so near shore would be highly unusual. The shark was six to seven feet, and Casey and his team knew what it was. It was a juvenile white shark, and on the next hook, the crew was presented with yet another juvenile white shark. The second fish was approximately the same size as the first, but two white sharks in the same location, on the same longline, right off a swimming beach? The men thought they had seen it all. Until they got to the third hooked fish, which was another white shark. The men realized something really wasn't right. Three large juvenile white sharks captured near shore? Three white sharks? Three sharks in general was unusual in itself, and three white sharks was unheard of! With each hooked fish, the scientists' astonishment grew. By the time they had heaved the single longline back to the boat, no fewer than ten juvenile (less than ten feet) great white sharks were sitting in the stern deck.

This extraordinary episode may have a correlate in 1916. One shark, besides the mako, that looks somewhat like a young white shark is the porbeagle shark. The porbeagle does not grow as large as the white and does have some distinct physical differences to the trained eye; most notably its teeth are not broadly triangular and heavily serrated like the whites. The porbeagle, however, is a strictly pelagic shark and has been implicated in only two attacks on man because it inhabits cold, deep waters. Shortly after the Matawan attacks, a *Chicago Tribune* headline read RAVENOUS PORBEAGLES DRIVEN TO BEACH BY HUNGER, IT IS BELIEVED. Might it not be possible that the shore fishermen or even the imperfect expert of 1916 misidentified an inshore school (admittedly unusual) of young white sharks as the better-known offshore porbeagles?

John G. Casey and Harold L. Pratt Jr., in a 1985 article, "Distribution of the White Shark, *Carcharodon carcharias,* in the Western North Atlantic", reported that over several years of recording, New Jersey and New York shark-fishing tournaments had reported a substantially higher percentage of white sharks caught compared to the percentage reported by the commercial shark-fishing industry in southern United States waters. Casey, however, pointed out that several factors should be taken into consideration when reviewing the data: (1) a disproportionate amount of fieldwork by biologists compared to other regions; (2) more intensive recreational and commercial fisheries in that region; (3) a closer working relationship between the biologist and fishermen in the mid-Atlantic bight who are aware of the scientists' interests in the white shark.

The white shark incident of August 1964 was certainly astounding, if not monumental, but the conclusions by Casey and Pratt on the general population of white sharks in that region were surprising, as well. The scientists concluded that the white shark was more abundant on the continental shelf between Cape Hatteras, North Carolina, and Cape Cod, Massachusetts, than in any other region of the western North Atlantic. The men also discovered that more young white sharks had been caught there than in any area of comparable size in the world. According to their diagram of white shark sightings/catches from 1800–1983, the highest density appears to be to the south of the New York bight, off southeastern Long Island, New York, and off the northern and central Jersey Shore. This temporal distribution density also corresponds to the months of June and July (e.g., July 1916).

On the topic of great whites and the Jersey Shore, I can't help but mention an incident that David Doubilet, the renowned *National Geographic* underwater photographer, has told regarding one encounter. Doubilet was raised in Elberon, New Jersey, and got his first academic taste of shark natural history from Jack Casey at Sandy Hook. Doubilet's early work at Sandy Hook, however, was not his first entrée to the world of sharks. He enjoyed snorkeling off the beach in front of his house and recalled discouraging a pesky shark from nibbling at his dive bag, which was filled with a catch of lobsters. As a teenager, he vividly recalls bodysurfing near his home and seeing a great white right off the beach.

In all fairness to New Jersey bathers, and in the spirit of scientific accuracy, the white shark is not plentiful off the Jersey Shore. It is not plentiful anywhere in the world. What New Jersey and Long Island waters may have (or had) is a pupping ground of whites and, therefore, a preponderance of juvenile white sharks. Yes, very large individuals are caught there, but fewer than the younger ones. It is a relative abundance as compared to other locations in the world, but I would not be surprised if certain locations around South Africa, like the southern and southwestern cape waters, could call the white something other than "rare." John Geiser, the *Asbury Park Press* fishing writer and author of the *Shore Catch,* reported that the New Jersey Bureau of Marine Fisheries survey finds only six-tenths of one percent of all shark-fishing trips (by anglers polled) resulted in the catch of a great white.

For comparative purposes that are obvious to this investigation, a look at the bull shark and the white in New Jersey is illuminating. Most authors and icthyologists agree that, although the bull is considered a local species, it does have straggler members (accessory individuals?) that stray from one region to another. Again, the Lake Nicaragua sharks may be an example of such straying. Individuals have been known to stray from southern waters to New York during summer months, but such an occurrence is not common. Jack Casey informed me that the bull shark is clearly more common than the white in waters south of Cape Hatteras, but it is much less common than the white north of the cape. In 1986, he estimated that three to four bull sharks or fewer are caught in New Jersey waters during the summer months, as opposed to an approximate average of twelve great whites.

The uncommon appearance of the bull shark in New Jersey waters may be a reason in itself to explain the rare events of 1916. For instance, on the rare occasion that a large, dangerous bull shark makes its way to New Jersey summer waters, attacks can occur. But why haven't 1916 rogue-type sequences occurred in Florida, where bulls are plentiful? Attacks in Florida are common, but they are rarely fatal; much more commonly they are attributed to sharks other than bulls. The rare life-threatening or fatal attack is, however, occasionally attributed to a bull shark.

An area of the United States that has seen fairly frequent serious attacks is the region known as the "red triangle." It is an area covering one hundred miles of northern and central California coast and was once considered the shark-attack capital of the world. The California white sharks are lured closer to shore by marine mammals, and these sharks, like most of the attackers, are larger whites than those identified in New Jersey. The California attackers, as in Australia, are adults whose diets include a high percentage of marine mammal. In 1994, however, researchers reported a precipitous decline in white shark populations along the California coast, and it is now a protected species. Between 1990 and 1995, California saw an average of four white shark attacks a year, with five in 1990 and 1993, with two fatalities. These numbers are compared to one attack in 1997 and 1998. Despite the decline in attack numbers, surfers still feel the weight of the waters they share with the declining white. Dr. Bill Rosenblatt, for example, a pain-management psychologist and prominent member of Surf Rider Foundation from New Jersey, has surfed the waters of Florida, Hawaii, New Jersey and California, and says he rarely thinks of sharks even with the occasional dorsal fin breaking water off the well-known surf break at the Eighth Avenue jetty in Loch Arbor, New Jersey. The only time he second guesses a paddle out because of potentially dangerous sharks is in northern California.

In New Jersey, with fewer small inshore marine mammals to munch on, the white sharks are usually smaller juveniles that rarely venture inside ten miles from shore. These juveniles feed mainly on small demersal fishes, including menhaden/moss bunker (*Brevoortia tyrannus*) and sea robins. These sharks almost seem as if they are consuming whatever they can find.

Even though the New Jersey/New York white shark population is mostly made up of juveniles, it is not uncommon to capture large adults that are possibly pupping their young. In a 1985 publication on the subject, Casey and Pratt reported 215 juvenile whites in this area and 66 adults, of which 8 were paired. The pairs were in the same general location (off Montauk, New York). In August 1986, twenty-five miles south of Montauk, Long Island, famed shark hunter Frank Mundus and Donnie Braddick unofficially broke the rod/reel record for a great white set by Alfred Dean of Australia. In 1959, Dean used porpoise as bait to catch a 2,664-pound white shark. Mundus and Braddick had tied their vessel to a whale carcass and pulled in a 3,500-pound, seventeen-foot mammoth white shark. A week later, a 2,600-pound white was snared in the same area. Both sharks had been interested in the whale carcass. Unfortunately for Baddick and Mundus, the International Game Fishing Association disqualified their catch because of the regulation that prohibits the use of mammal parts for fish catches. Alfred Dean's record still stands because it predated the rule on mammalian bait.

Ten juvenile white sharks caught by Jack Casey and crew off Sandy Hook, New Jersey in 1964.

CHAPTER 11

Closing the Case, for Now

From the beginning of this investigation into the infamous events of 1916, the very process of accumulating the evidence and data has brought me pleasure and satisfaction. My intention, however, was not only to pursue information that would help in a quest for a culprit or solve the mysteries related to its causation, but to provide readers and potential researchers with the material they could utilize to form their own opinions. Considering all that has been attested to and discussed, I would hope that we are all a bit closer to delineating the most probable identity of the shark or sharks responsible for the 1916 attacks. The puzzling and unprecedented events of eighty-five years ago are certainly frustrating to the analytical mind, especially when conflicts and contradictions are discovered in the records. Some have said that the hard "evidence is long gone" while others say the "evidence remains" for the serious student of shark attacks to ponder. Mary Batton suggested that a DNA analysis of the stomach contents from the Raritan Bay great white could have conclusively proved or disproved a connection between that shark and the 1916 victims. The process would compare the tissue or bone DNA from the stomach remnants to tissue samples from among the five 1916 victims. Unfortunately, that white shark's stomach remains were disposed of long ago, or they are at least unobtainable at the current time. I requested a formal search for those bones and the search came up empty. There was not even a log or catalog entry for such items. I am aware that Dr. Frederic Lucas, as an osteologist, received many samples of unidentified bones and fossils from assorted public sources over the years, and I am a bit dismayed that the Schleisser shark stomach remains are unavailable. We are, therefore, forced to defer to more tenuous forms of evidence.

Dr. Lucas's assertions that the long bone from the Schleisser shark was not a shinbone, and the contention that the Matawan Creek events were freshwater attacks, has swayed many modern theorists to exonerate the Schleisser white shark from involvement in any of the five attacks. Dr. Lucas's opinion about the controversial bone even caught the attention of *Scientific American* in September 1916. In that edition, the author noted that the bone identified by Lucas as being from a forearm likely came from an attack on a corpse, not a living person. The magazine writer also stated that it is impossible, therefore, to pin the attacks on any one shark (specifically the Raritan Bay white shark) "unless it is caught in the act."

The importance of those bones in the evaluation of these attacks is obviously paramount, and if Dr. Lucas were alive today, I would certainly approach him with a few choice questions. Namely, if he realized that the bony remains were the only indisputable elements of evidence and part of the few critical pieces of this shark-attack puzzle, why were they not made available for independent examination? Today, some scientists challenge the anthropological findings relating to prehistoric man simply because only a select group of anthropologists have access to the fossils. Also, in an institution like the American Museum, where prehistoric and antique objects and organic material are preserved and respected for perpetuity, why were these items of evidence apparently not protected or at least catalogued in the same fashion? Perhaps there is a simple and legitimate answer to these questions, but regardless of the loss of those bone sections and Mary Batten's thoughts on the lost opportunity for DNA analysis, the most convincing potential evidence to determine the species responsible for the attacks may be buried in the cemeteries of New Jersey. Among the victims' remains, most notably on Lester Stillwell in Matawan, we may find a microscopic tooth fragment. On this topic, I have spoken to Dr. Gordon Hubbell, a Miami veterinarian and shark-tooth maven. I have also spoken with Marie Levine of the Shark Research Institute (SRI) about her experience with such examinations. We all agree that finding a tooth fragment would provide the species' identification/confirmation. We could even match a Matawan victim's tooth fragment with that of one found in Charles Bruder or Charles Vansant. To exhume any one of the 1916 victims, however, would likely be impossible because of the current strict New Jersey and Pennsylvania exhumation criteria. In any event, the prospect alone of exhumation is probably outside the realm of research ethics or responsible scientific curiosity. Even without the evidence that may have gone to rest with the dead, I still believe that the data examined has tipped the scale of probability toward some specific designations.

It is certainly within reason for a student of the 1916 case to conclude that any one of the top three world-attack culprits, or any combination of the three, could be pinned for the recorded attacks. Perhaps you may be inclined to go along with the theory initiated by Richard Ellis or alluded to by George Llano that a bull shark is the most logical assailant at the Matawan Creek. If you believe that the bull was the attacker at the creek, then maybe you would assign the ocean attacks to the Raritan Bay white shark or some other unidentified shark like a tiger shark or even the same bull shark that appeared at Matawan. You may even be inclined to put credence into the impact of the "shark year" theory and consider that five different sharks acted independently.

Answers may also lie in the attacking shark's (or sharks') inclination to feed on baby sharks. As we know, 1916 saw an explosion of pupping sandbar sharks in the bays. At least one of the large sharks caught by Paul Tarnow in the Keyport Bay had swallowed fourteen baby sharks before it got caught up in a pound net.

Simpler yet, look to the heat wave of that season and the burgeoning New Jersey summer population, replete with its own U.S. president. Over and over again, modern statisticians find that population density is the greatest positive influence on shark attack. To some, perhaps the 1916 mystery will have been satisfactorily solved with these strong influences. They all push toward a reasonable degree of probability.

However, the fact that the entire 1916 sequence of attacks is generally unique and unprecedented leads me to believe that the cause must be unusual as well. With the peculiarity of the events somewhat understood and appreciated, application of the finer points of shark behavior and the obvious and subtle environmental features of the Jersey Shore can bring the culprit(s) and its actions into the realm of the possible as opposed to some unlikely, conveniently explained, unfathomable, supernatural, or supersensational phenomena.

As to the general influences that contributed to the attacks, the one I consider most influential would be the obvious one; the heat wave combined with the tremendous human population flurry during that early July period. Don't forget that the vintage bathing suits (tights) contributed to the sensation of overheating along with the ambient temperature. Interestingly, even though there were hundreds of people in the water and on the beach in Spring Lake during the Bruder attack, Bruder, like Vansant, was not swimming near any other bathers. The general increase in human aquatic activity must be counted into the equation, but the fact that these two men were solitary targets increased their odds of being singled out by a shark. Perhaps the 1916 shark(s) thought a new marine mammal had moved into the neighborhood. Even though sharks were said to be around more than usual in 1917, the migratory patterns of people to the Jersey Shore were diminished because of our April entrance into World War I.

One other potentially influential factor that was specific to the era was the bathing attire, or in the Matawan Creek, the lack thereof. Contrasting colors or shades have long been known to influence mistaken-identity attacks, and what better to create such a contrast than the black swim tights on Charles Vansant, Charles Bruder, Joseph Dunn and Stanley Fisher? As mentioned, some native pearl divers cover the white soles of their feet to disguise the contrast. The boys at the Matawan Creek must have had splendid tan lines acquired from afternoons playing baseball. The victims, especially the men, may have even appeared to be marine mammals or even "shell-less turtles."

A last thought on the apparent influence of the 1916 heat wave and the attacks should be stressed. We have already discussed the fact that the theorists were proven wrong in an assumption that the Gulf Stream went off course, but we have not stressed enough the possibility of the water temperature being abnormal for an average July. The interesting fact is, the ocean water temperature in Beach Haven was sixty-eight degrees Fahrenheit, which was somewhat cool for a temperate region attack. And although the water temperature was possibly on the high side (low seventies) just days or hours before the Bruder attack, the wind had switched from out of the north to out of the south by the afternoon, and the bathing regions were *colder* than usual. This cooler water temperature was said to be the result of a net northward water current. All old salts, and even some young salts, know that a south wind (from the south) creates a numbing summertime chill in the water. This change in water temperature can come on within hours. That temperature drop would discourage bathers from entering the water or prevent them from venturing far out. One analyst who was a contemporary of the 1916 attacks felt that in the case of Bruder and Vansant, the attacks would not have occurred if more people had been in the water. Such a contention actually has concordance with the modern view that swimming alone is more attack-provoking than swimming in a crowd. As one prophetic reporter wrote: "When a man goes out alone, far into the breakers, he is undoubtedly a shining mark for a hungry shark." What shark, among the top three attackers, would prefer cooler water temperatures to the warmer seas? The white shark, of course.

As I have mentioned, tropical species of fish do venture or stray to New Jersey and New York waters during the summer. I strongly doubt, however, that a stray tropical species (i.e., a bull shark or tiger shark) would arrive on the Jersey Shore in *early* summer.

In regard to other environmentally based attractive agents, such as horse carcasses or fish entrails, etc., I had mentioned earlier that sewage disposal in developing countries has long been known as a lure to sharks. Since 1916 was one of the first years that raw sewage was extruded into the sea in New Jersey, we may have yet another significant general factor to influence the mysterious attacks. This sewage factor, however, becomes an even more significant player at the Matawan Creek. Dr. Lucas always wondered what was so intriguing about that Matawan Creek to give that shark the inclination to poke its head upstream. Only one day after the attacks, it was speculated that the new sewer outfall into the creek may have attracted the killer shark.

Other than the general factors mentioned above, additional evidence relates to particular species of sharks. I realize that the bull shark's freshwater capabilities and actions have made it a popular species to pinpoint for the Matawan attacks, and I also concede the attractive potential connection between young sandbar sharks and the bull's tendency to munch on them. The argument that one or more of the Tarnow sharks (or the Cottrell shark) was a bull shark is strengthened by the fact that at least one of them had baby sharks in its stomach. Still, as I have mentioned, I believe the large Tarnow shark to be a dusky shark. The review of evidence that implicates the bull most indelibly, however, hinges on the premise that the Matawan attacks took

place in a freshwater location. The creek attacks were said to have occurred "11 to 16 miles distant from the open sea." Without that piece
of the bull shark contention, however, the bull shark theory is as probable (or improbable) as a tiger shark theory. Would not the bull shark
theory crumble or take a serious hit if the Matawan Creek were not
freshwater?

My scrutiny of the legitimacy of the bull shark proposal is not meant
to contradict any author's logic or knowledge of the species involved.
However, a closer look at the topography and geography of Monmouth County, New Jersey, namely the Matawan Creek and vicinity,
does raise a weakness in the bull shark premise.

In 1987, in my preliminary conclusions about the culprits involved,
I stated "general probabilities" in attaching a guilty party to each of
the attacks. Presently, after having further reviewed wound characteristics and other environmental and behavioral patterns, I feel comfortable in asserting more specific and definitive designations.

Almost every researcher who proposes or supports the bull shark as
the guilty Matawan Creek marauder begins the argument by saying,
"The only large shark to have any inclination to enter freshwater is the
bull shark." With this statement I have no objection, except to define
it further by specifying that the bull shark is the only large shark that is
inclined to continue meandering up freshwater after it has penetrated
the brackish initial portions of a waterway. The problem is, I don't see
the freshwater part of this statement as being relevant to the Matawan
Creek incident.

Earlier, we discussed the complex web of life that exists, almost hid-
den and unappreciated, in the world of estuaries and salt marshes.
These wetlands represent the interface links between the ocean and
the land or the ocean and the freshwater bodies. Just west of Sandy
Hook, New Jersey (just inside the hook), you will find a bay between
the mainland of Belford/Atlantic Highland's elevation. This is the
Sandy Hook Bay, and it represents the end point for boats exiting the
Shrewsbury and Navesink Rivers enroute to the open ocean. Those

rivers are so intercommunicative with the sea that blue claw crabs, bluefish, striped bass, bottlenose dolphins, and seals make regular appearances there. Moving a bit north of the Sandy Hook Bay, you will find the much larger Raritan Bay, which generally represents a wedge-shaped cutout of land occupied by Atlantic ocean water and whatever freshwater that drains from its western portions. The Sandy Hook Bay eventually feeds into the Raritan River (farther northwest). At the southwest terminus of the Raritan Bay, you will notice a much smaller but distinct bay known as the Keyport Bay or Keyport Harbor. Looking at the Keyport Bay, you would conclude that it is generally made up of the same waters that the Raritan Bay is composed of. The Keyport Bay is a bay within a bigger bay, similar to the Mattapoisett area of Buzzards Bay. Both the Raritan and Keyport Bays directly face on the Atlantic Ocean. The Keyport Bay used to be a popular ferry dock for commuters heading to Manhattan off West Front Street.

The Keyport and Raritan Bay areas are collectively known as the bayshore, and it is the westward (landward) side of the Keyport Bay that creates the mouth of the Matawan Creek. As you will recall, the creek is a progressively narrowing waterway that was so navigable (up until the 1953 fill project) that it saw colonial privateers and New York–bound steamers making dock landings all the way up to the Matawan train trestle. You may also notice from the 1916 Wykoff dock photographs that the creek was well maintained and dredged to provide a fairly significant clearance, both for depth and width. None of this is surprising, since the Matawan Creek is not a remote inland freshwater river or stream, it is a typical salt creek or tidal river.

Today, if you were to venture to the creek side near where the attacks occurred, you would be immediately impressed by the standard eelgrass and mudflats that make up the salt marsh. I have sunk many a time (with television crews, no less) into the quicksands that now line the low tide banks of the creek. Prior to sinking several feet into the mud, you would notice hundreds of fiddler crabs scurrying back to their tunnels, as well as many oysters shells. The salt or brackish water inhabitants, both invertebrates and finned fishes, do convince you of the "saltish" water presence. A self-conducted hydrometer study additionally revealed that the Matawan Creek, even today, is a standard temperate tidal creek.

Even more compelling than the creek's composition and general physical layout is an actual analysis of its distance from salty open ocean/bay water. All of the early literature and newspaper accounts report that Matawan is at least miles from the Atlantic Ocean. Unfortunately, this statement gives the impression that the location of the attacks is a long way from an area where sharks would naturally exist. This could not be further from the truth. That sandbar sharks, bull sharks, *and* white sharks are known to inhabit bays is now generally understood. And wouldn't the white shark be the most likely dangerous shark to inhabit the northeastern bays? The presence of the white shark caught by Schleisser in the Raritan Bay, the white sharks spotted in Buzzards Bay, the bay attacks in northern California, South Africa, and South Australia, and the population/distribution studies performed by Jack Casey et al. all confirm the white shark's ability and tendency to enter shallow brackish bays.

In actuality, the Matawan Creek attacks occurred one and a half miles distant from the mouth formed at the Keyport Bay. For twenty-five years, the bull shark has been considered the likely culprit at Matawan because of its ability to negotiate one and a half miles up a salt creek, at high tide. But the shark was obviously not content to continue up the more dilute/fresh aspects of the creek, especially with the peak of high tide approaching. Instead, it ventured back out (east) toward Joseph Dunn very quickly.

A few other general reasons point to a white shark as perpetrator. Because the boys in Matawan claim to have seen the white-shaded underside and gleaming teeth of the attacker on Lester Stillwell, the shark actually could have been peering above the murky water level at

Lester, a trait unique to the white shark. Also, most if not all of the 1916 attacks occurred with the attacker coming initially from behind and below, which is also the standard method by which the stealth-prone white shark works. Additionally, at Beach Haven, the witnesses claim to have seen a bluish-gray shark attack Vansant. Captain Cottrell, on the Matawan bridge, noticed the shark to be "dark." Joseph Dunn reported seeing a "black shape" under him. Such references are more specific for a great white than they would be for a bull shark, which is normally on the pale-gray side. The Schleisser shark was variously reported as blue and *dark*-dull blue. The tiger shark can also have a bluish tinge.

One of the old initial theories presented in the 1916 newspapers held that overfishing might be the cause of the shark's new hunger. Specifically, it was overfishing of menhaden. It was a reported fact from the fisheries' officials that bluefish, weakfish, and herring were very scarce that season. The prices for these fish were the highest they had been for thirty-five years. The menhaden that was commonly used (and still is) to fill lobster pots and create chum for bluefishing were said to be the scarcest of all the fish and at its worst point in twenty-five years. I'm not certain whether menhaden were truly overfished to some point of depletion, but if they were on a decline, it could have affected a frequent and abundant food staple of the local juvenile white shark population. Stomach contents from white sharks caught in the New York bight from 1961–65 revealed that, of the food contents identified, 24.1 percent was menhaden.

Any closing opinions on the 1916 attacks would not be complete without a final look at the rogue possibility. We may even be surprised to find our answers about the causes of the attacks and the culprit(s) in the attacks from this very angle of the analysis. I will avoid committing to any assertion that the 1916 attacker was a man-crazed shark that tuned its feeding objectives toward the pursuit of human flesh. What I will commit to is the notion that one shark did indeed perpetrate the five attacks within eleven sunsets.

A look at the Matawan attacker should tell us immediately that that shark, which for all realistic purposes was one shark, was a very unusual character. The Matawan/Cliffwood attack event, if perpetrated by a single shark, is, alone, the most fantastic multiple attack sequence in history. The shark, almost intentionally, killed and gnawed at the body of Lester Stillwell. (Such a display is markedly similar to the actions of the nine-foot white shark that repeatedly mauled Theo Klein for twenty minutes in South Africa in 1971.) This shark, after taking Lester's life, went on to devour two-thirds of Stanley Fisher's right thigh and, after being subjected to flying boat oars, traveled swiftly eastward and audaciously clamped down on the left leg of Joseph Dunn. If that shark did not display characteristics of Stewart Springer's accessory population, I don't know what does.

The attacks on Charles Bruder and Lester Stillwell were marked by the highly unusual behavior of a shark that struck back on its weakening, frantic prey *multiple* times. The attacks on Vansant and Fisher were so similar in anatomic location, intensity, and general size of the victims that both men expired in similar manners and time frames.

It is easily within the realm of possibility that Joseph Dunn could have been pulled from the water that day with one of the exact lower extremity wounds that Charles Bruder experienced. Joseph was spared that gruesome fate because of the heroic efforts of his companions. Additionally, the companions may not have been the only stimulus to turn off the shark's feeding drive at that moment. It almost seemed as if the shark wanted to perform the same type of "flesh stripping" on Joe Dunn's leg that it had done on Bruder's lower, narrow bones. In Dunn's case, however, the shark found that there was not much meat to strip, and the shark's leverage could have even been affected by the smaller bite mass.

The timing and distances between the 1916 killings also links to the nature of the attacker. The attacks in Beach Haven and Spring Lake occurred five days and forty-five miles apart, while the Matawan events occurred forty-nine miles distant (it's thirty miles as the crow flies) from Spring Lake and six days later. These intervals and distances reveal a shark that is generally moving at the same speed (northward) and is preying on large prospective food items at approximately the same interval of time and distance.

Not only was there concordance in the time interval based on days, but the shark attacks also took place at very similar times on those given days, all in the late afternoon. The Spring Lake attack and the Matawan attacks also took place within an hour of each other (around two o'clock). One may say that shark attacks simply take place predominantly in the late afternoon, but we still must not dismiss such a potentially informative pattern. Additionally, we must not forget that it is the great white shark that is the predatory shark that most relies on daylight for feeding.

The functional similarities between the attack locations is also interesting. Assume for a moment, although not critical to the case, that the solitary shark involved in the attacks (or even multiple attackers) was hunting down the sandbar pups of the bays. There would be no better locale for such a predilection than the Beach Haven region, which is very close to the Great Bay of Little Egg inlet and the long Barnegat Bay. At the Spring Lake spot, the Manasquan inlet and the Shark River estuary are just south and to the north, respectively. The Matawan Creek, as we know, is a tributary of the Keyport and Raritan Bays.

When that shark got north as far as Bayhead, near the Manasquan inlet (just south of Spring Lake), it had another strong influence to direct it on a northward path. In Bayhead, New Jersey, a distinct northward current originates that travels the stretch of the remaining shore. It is referred to as the littoral drift and was discovered by chance when early shore inhabitants noticed that shipwreck victims always floated northward in Monmouth County (assuming the storm was not a northeaster). A quick glance at any aerial shot of Monmouth County's shoreline (also known as the jetty county) would reveal sand accumulations at the southern side of the sea jetties and a paucity of

sand on the northern side. This sand movement trend is a daily reminder of the littoral drift. We have also heard that a steady and heavy
south wind (northward) was creating a chilly early July sea. We also
know that sharks often prefer to follow drifts, as do baitfish. The
sharks also tend to follow inward currents and inward, rising tides. In
other words, the shark from Beach Haven followed the shoreline
northward along the southern coast and, finding it could not gain access to the Barnegat Bay on the northern edge of the eighteen-mile
Long Beach Island stretch, proceeded north toward Spring Lake and
the bays in that vicinity. On the way, it got caught up in the littoral drift
(which begins at Bayhead—about seven miles south of Spring Lake)
and, after snapping Charles Bruder's legs, it ventured north until the
drift released it at Sandy Hook. At Sandy Hook, it turned left (west)
and into the inviting pup-rich Raritan and Keyport Bays. As the *Star-
Eagle* wrote just after the second attack: "The currents have been running northward for nearly a week, and it is figured that the shark
would go along with the trend of the current rather than struggle to
fight his way back to the south. It is certain that he persists in hugging
the shore as he has done since he made his first appearance at Beach
Haven last Saturday."

The solitary "roguelike" movements and actions of the 1916 shark
are certainly plausible, but what about a substantial link between the
attacks and the Schleisser white shark? Dr. Frederic Lucas determined
the bones in the shark's stomach to represent not a "boy's eleven inch
shinbone" but the radius and ulna (forearm bones) and a badly shattered anterior (front) rib fragment of a robust predeceased man.
These bones have, as stressed, mysteriously disappeared (or are at least
unaccounted for) from the American Museum, which traditionally
saves all important artifacts. A simple explanation for the bones' disap-
pearance could be Lucas's "obligation" to return them to Schleisser.
Since the physicians and Dr. Lucas did not have the luxury of CAT
scans, MRIs, or other sophisticated modern medical equipment to de-

lineate the bone identity of these incomplete portions, it would have been at least more thorough and objective to allow another anatomist, osteologist, physical anthropologist, archaeologist, or an orthopedist a chance to examine the bony remains. Lucas and the physicians simply utilized their powers of observation and their knowledge of comparative bones of similar length, width, and contour.

In the fall of 2000 I received a very unexpected e-mail photo from South African attack expert, Marie Levine. The old photograph depicted an intact right forearm and hand and two dissected sharks. Levine forwarded the photo, which she received from a Florida man who claimed that his grandfather snapped the shot and that the picture depicted the 1916 killer shark and its stomach contents. Levine felt that the photo represented an authentic attack-related limb because of its defense wounds. When I examined the photo I immediately wondered whether it represented the firm proof that Lucas was correct in his assertions about the unrelated "forearm bones" found in the Raritan Bay white shark. But why was this flesh-covered arm unreported with the Schleisser dissection? Was this New Jersey's equivalent of the "shark arm murder case"? When I forwarded the photo to George Burgess in Florida, I got my answer within minutes. Burgess recognized the arm immediately as the remnant of an offshore plane crash/shark attack victim which was documented in *Shark! Shark!* by W. E. Young (1933). The 1916 case was still open, therefore, and only bones were available for examination.

Unfortunately, from a forensic standpoint, Lucas and the physicians had to compare these incomplete bone portions to their knowledge of *whole* bones. One aspect of this effort that would be vital, besides general size and shape, would be to delineate the presence and location of tuberosities. Tuberosities are the normal bony growths or elevations on the outer surface of bone that relate to attachments of tendons. On the radius (at the proximal end), for example, a tuberosity exists for the insertion of the biceps tendon and for the tibia (shinbone) it occurs toward the knee for the patella (kneecap) tendon.

The bones would also possess depressions that would relate to certain muscular locations. These locations would be very bone specific and helpful, *if* the bone fragment were whole enough to include such contours or bony growth.

If Lucas was correct, then the Raritan Bay shark likely possessed at least one bone portion that was not taken from the known 1916 victims. None of them received forearm bone loss. The radius and ulna, however, could easily be confused with a clavicle (collar bone) or fibula (lateral lower leg bone) which Lester Stillwell likely had fractured. The doctors at the dock may have been influenced to call such a narrow bone the "shinbone of a boy," assuming that either Joseph Dunn or Lester Stillwell lost such a part. The physicians also knew it was too narrow to represent the tibia (shinbone) of a man. Additionally, the fibula bones of Bruder's lower legs (at least the fibula) can also be thrown into the confusion for the same reasons. In other words, perhaps this "radius or ulna of a robust man" was actually part of the fibula of a boy or of a Swiss bellhop.

The shattered anterior rib fragment also presents some questions, but because of similar potential confusion, its identity is also in question. Charles Bruder, however, did receive an anterior "abdominal" wound about the size of an "apple," which could, if high enough, explain the excision of this unaccounted rib fragment. While Colonel Schauffler simply identified it as an "abdominal" wound, the hotel physicians noted that it was "below the rib," meaning it was near enough the ribs to point out. The lower fibs can be deceptively low as they approach the lower flank. The Schleisser white shark could have spiked its lower, angled, more penetrating bottom row teeth into the right portion of Bruder's upper abdomen and "dug" upward to excise and shatter a rib section. When the tissue retracted downward after the strike, it would appear just below the ribs to the physicians.

Another interesting aspect of the white's stomach contents relates to the fact that the weight of the human flesh (with the bones) was said to be fifteen pounds. The doctor at the Matawan Creek who examined Stanley Fisher's immense thigh wound estimated that he was missing ten pounds of flesh.

If a reader were inclined to accept the notion that the stomach contents of the Raritan Bay white shark were human remains, which were unassociated with any of the known 1916 victims, he or she would also be making another automatic assumption. Namely, during the spectacular shark-attack spree, some unaccounted-for "robust man" drowned or was killed at sea, and while no one spoke up to say he was missing, a juvenile white shark happened by and had a snack on this man's arm and rib section. That's quite an assumption.

Assuming, perhaps, that the flesh from Charles Bruder would be digested to an unrecognizable state by July 14, (quicker than bone), my guess is that the Raritan Bay white's stomach contents consisted mainly of some of Charles Bruder's bone remains and some of the flesh from Fisher and Stillwell/Dunn (ten pounds from Fisher and five pounds from the youngsters and Bruder bones). Regardless of the gory and frustrating conjecture on the bone and flesh remains, one thing was for sure: the Raritan Bay shark certainly had strange eating habits if it included flesh, limb, and rib parts of human beings.

Another interesting fact about this Schleisser white shark has to do with its capture. It was said to have been caught in a drift net, with an eight-foot motorboat, near the mouth of the Matawan Creek. What does this tell us about that shark? By being caught in a drift net, which was no more than six feet below the surface, the shark was either attracted to the boat, attracted to the small fish that may have been trapped in the net, or was simply lumbering in the shallows and didn't/couldn't get out of the way. Either way, it sounds an awful lot like those strange accessory sharks of the Springer theory, which are "easily caught and attracted to boats." It even sounds like one of the partially disabled rogues we've seen of other species in the animal kingdom. Additionally, even though white sharks are known to inhabit shallow bay areas, they don't necessarily make a daily habit of it. The white sharks of New Jersey normally come no closer than ten miles off-shore. This shark, then, was not only in a suspicious area in relation to the Matawan tragedies (with human remains in its stomach, no less), it was in an unusual location for that species. Recall that it rained heavily the night before the morning that Lester's corpse was recovered. Sharks like the blue shark normally avoid nearshore migration because of the lowered beach line salinity. This Raritan Bay white shark, however, was captured in the bay just hours after freshwater dilution. There is little doubt, therefore, that it would certainly be physiologically capable of surviving a mile-and-a-half nasty sojourn up the high-tide heat wave–concentrated Matawan Creek of two days before. This white shark, perhaps dazed after its tumultuous creek visit, would have made its way out into the Keyport Bay, and eventually headed slightly *northward* again by Friday, July 14. On that morning, Schleisser and John Murphy happened by with their drift net.

References to body lengths and the size of the Schleisser shark are certainly interesting. The shark at Beach Haven was said to be nine feet long, and even though the witnesses at Spring Lake didn't see much more than a dorsal fin, they estimated it to be nine feet in length. The frantic people at the dockside in Matawan thought the shark to be nine feet long as well. Either these people knew of estimates from earlier accounts or they were great at uniformly estimating (overestimating?) shark length. Knowing what we know about witness exaggeration from the Mote Marine Laboratory trials, it would probably be safe to say that the witnesses of such events could easily have added one and a half feet to their actual visual assessment (or to the accurate size of the shark). To make matters worse, none of the "nine-foot estimates" was presented as a direct quotation from any one person. It's as if the writer recorded the buzz in the air or the buzz in his pen. But who could we look to for an accurate, credible, objective, unemotional length description for the creek attacker?

We know that Captain Cottrell, a sturdy man of the sea, reported the shark's length to be eight feet. Cottrell's bridge view of the shark was certainly optimal, but perhaps some may think that his estimate could have been blurred by the excitement. In a rare account caught by a dedicated reporter for a Jersey City newspaper, we do know of another "bird's eye" observation of the Matawan Creek shark, besides Captain Cottrell's. A motorboat filled with teenage boys was on the water at the time of the shark's entrance into the creek. The boat's operators got a good look at the shark just after Cottrell did. In fact, in later testimony, Albert O'Hara volunteered information that he and the boys were warned that a shark was in the creek prior to the incident that took the life of Lester Stillwell. O'Hara mentioned that a group of teenage boys came by in a motorboat, but Lester and the gang thought the warning was a joke. In that boat were three eighteen-year-old boys named Harold Conover, Ralph Gall, and John Tassini. Tassini reported that they saw the shark near the Matawan drawbridge at 1:30 P.M., and it appeared to him to be not much greater that six feet long. The Resort Edition of the July 13 *Asbury Park Press* subtitles one article: 7-FOOT FISH SEEN CRUISING UP THE RIVER. The length of Schleisser's catch was seven and a half feet, not a bad size to correspond with information from the more credible witnesses.

When we consider the reliable witness testimony pointing to a "7-foot-plus" creek shark as opposed to a "9-foot-plus" creek shark, we may also support such a contention by recalling the assertions made by Colonel Schauffler and Coroner Fay when they reported the Bruder and Stillwell attackers to be "small sharks." These men came to such conclusions because, in their examination of the wounds, they observed the true bite width of the shark (the shark did not utilize rotary-side-to-side excision methods on these particular victims). Charles Bruder had clear tooth scratches on his abdomen and Stillwell had single-bite avulsions to several locations of his corpse. An estimate of such a bite width, especially as it relates to an eleven-year-old boy's frame, certainly matches the 9.5 inch estimate for the Schleisser shark's jaw size.

I have put great weight and sentimentality into the testimony I've received from witnesses of the 1916 events. At this juncture, I would like to stress a few items of testimony that mean a great deal to this critical phase of the analysis. Almost every old-timer, male or female, child or adult at the time of the action, told me that he or she distinctly recalled it being "high tide" during the Matawan attack scenario. This fact, again, stresses this particular shark's tendency to follow a current and enter a brackish system that was not only wide and deep enough, but was adding seawater.

One of the most treasured pieces of audiotaped testimony that I possess is that of George "Red" Burlew. As touched on earlier, Red Burlew was not only in the thick of the search for Lester Stillwell's corpse with ill-fated Stanley Fisher, but he also became one of the finest and most sought after big-game fishing captains in later years. His business trail measured from Keyport to Brielle, New Jersey, then finally to the east coast of Florida. His objectivity and knowledge as a creek witness cannot be overstated. The astonishing clarity and coherence of Captain Burlew, who was in his mid-nineties during the telephone interview, was a tribute to his innate strength and longevity. Even in that late stage of his fulfilling life, he provided incredible specifics to the story, including a description of the creek attacker. Let us not forget that Burlew spent practically his entire life looking over the gunnel of his fishing vessels to eye the catches that were reeled to the side of the boat. A seasoned captain gains a sixth sense in his ability to glance at the splashing, struggling fish, and make precise identifications and weight estimates about the catch. It is with that in mind that I value the motion-picture-like memory of Captain Burlew. In his riveting words: "I was born in Keyport, and that creek at high water you could come up in the rowboat, ya know. That shark came up, and at low water he couldn't get out. It was the only shark I saw up there. He must have weighed about 300 to 325 [pounds]." As we know, the Schleisser white shark was 350 pounds. On asking Burlew if he thought a great white was the culprit, his response was, "It had to be."

In my mind, and hopefully without much imagination, I believe that if the Schleisser shark were embroiled in a courtroom case as the indicted killer in the entire 1916 attack saga, it would not get even a plea-bargain deal. For those apt to consider a bull shark, a large sandbar shark, a large dusky shark, and perhaps a tiger shark: imagine them or any of the sharks caught near the Matawan Creek to be standing in a witness lineup. Next to those sharks you would have the Raritan Bay white shark standing with its human stomach contents displayed in front of it. Which would you choose? How many other sharks during the sweeping Shark Hunt General had human remains in their stomachs, were caught near the scene of the last attacks, and had a species reputation to match the "crime"? Only one that we know for sure. That white shark was apprehended with a smoking gun right near the sight of the killings.

Most convincing, perhaps most overused, and even most belittled, is the explanation that the attacks stopped after the white shark was caught. Yes, opponents of the Raritan Bay white shark theory say this either proves nothing or is simply a coincidence. The fact is, the attacks did stop. Whether the attacks were perpetrated by the Raritan white shark or not, the ferocity and frequency of the attacks would imply that the attacks should have continued. If the shark were still on the loose, there was plenty more summer left on the Jersey Shore, and it certainly could have found a bather somewhere daring enough to swim outside the rickety nets. The culprit, however, must have encountered a dramatic roadblock. It could have been the shark caught by Captain Cottrell or the Tarnow shark, but the way I see it, the responsible shark was caught in the Raritan Bay, in a drift net, on the morning of July 14, by Michael Schleisser.

Fortunately, the Jersey Shore did return to calm after the 1916 blow. The atmosphere of golden innocence and well-mannered frivolity, however, never fully returned, even after the war to end all wars was over. Like the popular song by the Irish rock group The Cranberries says, "We've been fighting since 1916." At Ocean Grove, New Jersey, the town's Methodist Camp Meeting Association summarized the state's greatest problems of 1916 with great optimism for 1917 when they wrote:

> The development by the state and medical colleges of the successful prevention and cure of infantile paralysis apparently reduced the plague of the past summer to a level with other diseases under the control of regular practitioners. In the shark scare along the coast, the forms of defense against their attackers have been so thoroughly tested as to leave out of the question any necessity for further apprehension of danger. When to this is added the tests made by the government of the value of shark skins as a substitute for leather as affording remunerative employment for the fishermen along the coast, no one need be deterred from visiting the Grove because sharks have made their appearance upon our coast.

In October 1916, after the smoke had time to clear, Dr. John Nichols and Dr. Robert Murphy collaborated on another scientific publication for the Brooklyn Museum; this time the men composed the new look at sharks. Dr. Lucas did not contribute to the publication. In the paper, they discussed the tragedies and supported the view that the white shark in question was the killer. They also could not avoid alluding to the inaccurate assessment they had made in April 1916, when they spoke of the shark's limited jaw potential and capabilities for attack on the live human form. In this journal edition, Nichols wrote: "It must be admitted that deaths from shark bite within a short radius of New York City would seem to be one of those unaccountable happenings that take place from time to time to the confounding of savants and the justification of the wildest tradition."

By November 1916, President Wilson's campaign headquarters at Asbury Park counted up the election votes and gave Wilson the happy news of his reelection. Wilson could not celebrate too long, however, because, by then, German submarine activities were sending the message to the world that they were intent on winning a large-scale war. The Germans extended the war zone to cover the high seas and hoped that unrestricted U-boat warfare would win quick victory before the United States joined in to effectively intervene on the side of the Allies. In January 1917, the British intercepted a German message to Mexico, that indicated Germany had approached Mexico for an alliance in case of war with the United States. The Germans promised, as part of their payment, to help recover land that Mexico ceded to the United States after the Mexican War. In March, the Germans torpedoed several American merchant ships without warning. The next month, German submarines sank nine hundred thousand tons of Allied shipping, the all-time high for the war. Information of that kind persuaded the United States to declare war on Germany on April 6, 1917. When Nichols, Murphy, and Lucas penned the Brooklyn Museum article of April 1916, little did they realize that the world would change so dramatically in only one year.

If there ever turns out to be some fantastic link between the Germans and the shark attacks of 1916, one should never forget the story of Jonah. The "great fish" in the Jonah story was an obedient agent of God's purpose. In the 1916 attacks, the sharks or, specifically, the maligned Raritan Bay shark, would be seen as an innocent "obedient agent" of the Germans' purpose.

Dr. Nichols stayed on with the American Museum for the remainder of his long and illustrious ichthyological career and wrote close to a thousand publications on fishes, authored books on the fishes in the vicinity of New York and China, and even identified a shark's tooth fragment taken from the victim of an attack in Panama in 1927. Regardless of the mistakes and unpreparedness of the scientists called upon to make sense of the attacks in 1916, Nichols, at least in my opinion, should be applauded for figuring out the species of the perpetrator after the second attack. He nailed the great white before anyone else, before it was caught, and long before the species was mythologized in *Jaws*. He resided in Garden City, New York, until his death in the late 1960s. Dr. Lucas remained as director of the American Museum until his death in 1929. His influence on the museums of America was great and lasting, and he is probably best known for academic and administrative achievements and as an ornithologist rather than for his work as a field naturalist. Lucas had only three opportunities for field expeditions in his life, all three successful. He was not without a sense of humor during his days as museum director, and once was quoted as saying, "A museum is an institution for the preservation and display of objects that are of interest only to their owners. It is a place where paintings, bric-a-brac, trophies of the chase, etc., may be deposited whenever their owner wishes to have them stored temporarily without expense to himself." Robert Murphy was perhaps the most prolific scientist of the three. According to eminent biologist Ernst Mayr, Murphy would "work with iron self-discipline, no matter how strenuous the day, he recorded his daily experiences in considerable detail in a diary, an extraordinarily valuable record considering the drastic changes all of these places have experienced since then." Among his field endeavors as a pelgic bird specialist, Murphy traveled to Baja, California; Mexico; Peru; Ecuador, three times; the western Mediterranean; the archipelago of Las Perlas, off Panama; New Zealand; the subantarctic region, three times; and the Caribbean area, several times. His early and lifelong efforts also included the conservation of his home region's natural environment on Long Island. Murphy was best known for his published diary, which captured his 1912 expedition to South Georgia Island. Specimens from that expedition adorned the walls of the American Museum for more than fifty years, and, in 1967, Murphy's logbook excerpts were reissued as a *Time-Life* paperback. That book bears the legal disclaimer that could be most appropriate for this book as well: "All characters in this book are real persons, and are called, to the best of the author's belief, by real names. Any resemblance to fictional or other persons, living or dead, is purely coincidental." In 1970, three years before his death (at the ripe age of eighty-six), Murphy served as a representative of the National Science Foundation and biologist on the icebreaker *Glacier* in the Antarctic and revisited South Georgia Island, which he had last seen in 1912. Murphy never forgot his friendly relationship with the renowned Dr. Lucas, nor did he forget that it was Dr. Lucas who arranged for him to go on the Antarctic expedition fifty-eight years before. During that trip, Murphy repaid the favor and named a glacier off South Georgia Island for Dr. Lucas, and since then two mountains in the Antarctic have been named Murphy.

John Nichols and Robert Murphy could not have been more right in calling the summer of 1916 "one of those unaccountable happenings." To make sense of such "happenings," I hope that enthusiasts of this and other marine mysteries will not forget preexisting research and the vast experience garnered from assorted fields. If we assemble that data, perhaps we can promote a successful advancement in our powerful and useful body of knowledge.

In 1987, in *In Search of the "Jersey Man-Eater,"* I offered the impression that I had tapped every accessible source of information on the attacks, and that I felt a thorough job had been done to use that information to verify old accounts or gain valuable new evidence. With my involvement with research on no less than four videos on the subject, and this exhaustive approach, I must now admit that I have unearthed much more evidence postdating my earlier efforts of the 1980s. The new evidence has been brought to light in an effort to combine the scientific facets with the dramatic features of the story.

With information as the key to all progress, it is my hope that the steady advancements and efforts in worldwide shark research, the expanding shark-attack file, and exposure of personal collections or recollections may someday answer all the burning questions about the 1916 attacks. Such a culmination will satisfy *all* the theorists.

More important, the growing mass of data will, I hope, allow us to determine when a real shark-attack threat exists, depending on dynamic environmental conditions. The very fact that worldwide attacks are increasing means that science and governments cannot ignore the efforts to decipher the mysteries behind a shark's misplaced aggression.

In eighty-five short years we have come full circle with respect to our view of sharks. Most of us do not see the shark as a malevolent, chinless demon of the depths. We know the shark is a powerful predator, and we must respect the domain it calls home. We do not blame the bear that finds its way into our tent when we are camping or the bee that stings us during a picnic. The ocean, like all other areas of the wilderness and the natural world, is ours to enjoy, visit, and call home, but we must respect it and realize that it is also home to a great many species. If we were to place blame on anyone for the actions of an overzealous bear, perhaps those who have left trash around the campsite or encroached on the bear's natural habit should be called to answer. With the shark, let us not marvel or cower at how close some large, dangerous sharks have come toward shore unless we also make certain that longliners or other fishing vessels have not lured the pelagic species near bathing areas and that fishing or industrial trends have not somehow disrupted the entire marine ecosystem. Regardless of any human role in inadvertently influencing an encounter between the large predator and man, we can always blame mankind for one other thing: our gaps in research.

Perhaps our ultimate mastery in understanding shark-attack behavior will generate a precipitous decline in serious shark attacks. What would be even more productive would be for this shark-attack research to coincide with a powerful knowledge of how to re-create harmony in the seas and waterways and save this irreplaceable creature. In the process, perhaps our advances will tell us how to preserve or restore the fragile oceans, wetlands, and estuaries that have suffered the deadly blows of man. For our sake, I hope the day is not far off when we can correct our devastating environmental mistakes and learn the subtle behavioral habits and influences of underappreciated marine species and their niches.

Preservation of the marine and natural shoreline environment may not be the only challenge before us in the new millennium. My efforts to elaborate on the grand social setting of the post-Victorian New Jersey era were meant not only to give us a feel for life at that time, but to inspire preservation of all of the vintage buildings and structures that still remain along the Jersey Shore and the entire East Coast. If the search for the Jersey man-eater can promote realization of some of those objectives, then the tragedy of the summer of 1916 will not have been in vain.

Species of Sharks and Their Characteristics (a select list)

White shark, *Charcharodon carcharias* **(aka "Man Eater"),** the largest carnivorous shark. Its average adult size is 15 feet but it may reach up to 21 to 25 feet, and it is generally slate-gray to black above and white on its underside. The white feeds on fish, squid, seals, sea lions, whales, sea turtles, and other sharks. It has a worldwide distribution but is not common anywhere. It hunts during daylight hours and can be found in pelagic waters or within yards of the beach or docks. The white shark is blamed for three and a half times the number of attacks than either the tiger shark or the bull shark.

Tiger shark, *Galeocerdo cuvier,* is second only to the white shark as a danger to humans. It reaches 21 feet and its telltale stripes fade in adulthood as it takes on a sandy to dark gray upper body and an off-white belly. It is common in tropical and subtropical regions and can be found along the U.S. Atlantic coast in warm months. The tiger often appears close to shore and will eat almost anything that floats.

Bull shark, *Carcharhinus leucas,* is now blamed in many attacks where the shark was not positively identified. It reaches 11 feet in length and 400 pounds and its coloration varies from light gray to light brown. It is abundant in tropical waters and frequently enters the shallows and explores estuaries. Its tendency and ability to enter freshwater rivers has given it the regional names of Zambezi shark, Lake Nicaragua shark, and Ganges shark. In those regions, it is notorious for attacking people. It is occasionally found as far north as New York. They rarely appear on the surface prior to an attack.

Sand tiger shark, *Eugomphodus Taurus* **(aka Spanish shark),** grows to only six and a half feet but is considered a dangerous shark and has been known to be very nasty to aquarium handlers. The larger adults feed on marine mammals, and it often launches surprise attacks from below. The sand tiger is speckled in appearance and can be found in tropical waters and warm temperate waters during the summer months.

Blacktip shark, *Carcharhinus limbatus,* grows to eight feet and usually feeds on smaller fishes and has often been found in the stomachs of tiger sharks. It inhabits tropical waters worldwide and is likely a main culprit in the mistaken identity nonfatal attacks along Florida's east coast.

Great hammerhead shark, *Sphyrna mokarran,* has been reported to grow to twenty feet and has been implicated in many attacks on humans. It is common in the tropics and makes predictable mass migrations. Its size, aggressiveness, and appearance have helped its reputation as a dangerous shark.

Blue Shark, *Prionace glauca*, is known for its brilliant blue color and winglike pectoral fins. It is a long-distance traveler and reaches up to thirteen feet. It is generally an inquisitive deep-water shark but has been known to attack people and has been blamed for feeding frenzies on World War II shipwreck victims.

Shortfin mako shark, *Isurus oxyrinchus,* is likely the fastest swimming shark. Divers should use caution when this shark begins a figure-eight swim pattern. Its body is sleek and usually a deep purple on its topside. The shortfin mako has stiletto-type teeth and it is mainly a fish eater, but has been blamed for nonfatal attacks on swimmers. It reaches thirteen feet in length.

Sandbar shark, *Carcharhinus plumbeus* **(aka brown shark, ground shark, or blue-nose),** is brownish gray and reaches a length of seven and a half feet. It is the most common shark in New Jersey and Hawaiian waters, and it is also one of the most common sharks on the entire east coast of the U.S. Historically it has used the bays of New York and New Jersey as pupping grounds. It is probably of limited danger to man.

Dusky shark, *Carcharhinus obscurus,* is often mistaken for other sharks because of its standard appearance. It is pale gray above and white below. It is a common large shark in New Jersey. It grows to twelve feet and is the shark most frequently entangled in the South African anti-shark nets. Its bad reputation is probably unwarranted.

Oceanic whitetip, *Carcharhinus longimanus,* is possibly the most dangerous shark of the open seas. It was notorious for terrorizing torpedo stricken sailors during World War II. It is a warm-water shark and reaches a maximum size of thirteen feet.

Porbeagle shark, *Lamna nasus,* is a fast, sleek shark, with stiletto-type teeth similar to the mako. It is known as a fierce cold water predator in the North Atlantic. Statistically, it is not considered especially dangerous to humans because it avoids waters that are above 65 degrees Fahrenheit. It reaches thirteen feet.

Thresher shark, *Alopias vulpinus,* grows to twenty feet in length and is known for its long tail (caudal) fin. It does use its tail to corral fish but it is not a great threat to swimmers. If and when caught by a fisherman, it is likely that shark's first contact with man.

Whale shark, *Rhincodon typus,* is the largest shark and the largest fish in the sea. It grows to forty-five feet and perhaps up to sixty. Found in tropical waters and warm temperate regions, it provides us with the approximate living dimensions of the extinct megalodon (extinct ancestor of the white shark), but it is not considered dangerous as it lives on crustacea, tiny fishes, and plankton.

Timeline of Terror

1891 Hermann Oelrich's offer of a $500 reward for an authenticated temperate water shark attack appears in the *New York Sun.*

July 1914 World War I erupts in Europe.

1915 Heinrich Friedrich Albert perpetrates failed espionage mission for Austrian Embassy in New York.

Williamson brothers observe a shark feeding frenzy from a bathysphere for the first time.

Germans warn that ships will be sunk if an Allied flag of Britain is flown.

British liner *Lusitania,* with hundreds of Americans aboard, is sunk by a German torpedo.

April 24, 1916 American Museum of Natural History and Brooklyn Museum scientists publish an article on the great improbability of a true shark attack against man.

June 1916 100,000 U.S. National Guardsmen under the command of General John J. Pershing are sent to Mexico to capture bandit Pancho Villa.

Early July 1916   In Europe, Germany attacks at Verdun. Austria and Italy fight along the Isonzo River. Allies take the offensive along the Somme. Russia assaults the Poles in Galatia, and the fleets of Germany and Great Britain clash at Jutland.

New Jersey and national dignitaries arrive in Spring Lake, New Jersey.

The German submarine *Deutschland,* under the auspices of the North German Lloyd Steamship Line, cruises the Northeast U.S. Coast.

July 1, 1916   Charles Vansant is attacked and killed by a shark in Beach Haven, New Jersey.

July 6, 1916   Bell Captain Charles Bruder is attacked and killed by a shark at Spring Lake, New Jersey.

July 8, 1916   Steel netting is universally installed at Jersey Shore beaches.

July 10, 1916   Paul Tarnow captures a large shark on the Belford pound nets.

July 12, 1916   At Matawan Creek, Lester Stillwell and Stanley Fisher are attacked and killed by a shark, and Joseph Dunn is seriously mauled.

July 14, 1916   Lester Stillwell's corpse appears on the surface of Matawan Creek. The Raritan Bay white shark is caught. Ichthyologist John Nichols visits Matawan.

July 15, 1916   Shark protection conference is called at Asbury Park. A summit regarding the Mexican summit is considered for Asbury Park.

July 18, 1916   President Wilson decides that West Long Branch, New Jersey, will be his summer home and Asbury Park will be his summer White House. Captian Thomas Cottrell reports the capture of a large shark near the mouth of Matawan Creek.

Late July 1916   Cigar-bomb attacks by the "Dark Invader" are perpetrated on Northeast U.S. merchant vessels. At Bayonne, New Jersey, the Black Tom Island ammunition station is destroyed by German saboteurs. The *Deutschland* resurfaces in Connecticut.

August 1916   The New York/New Jersey border is guarded to prevent entrance of infantile paralysis–afflicted victims from New York City into the Garden State.

September 15, 1916   Joseph Dunn is released from St. Peter's Hospital, New Brunswick, New Jersey.

October 1916   American Museum and Brooklyn Museum scientists collaborate on a revision of their thoughts on shark attacks.

January 1917   German saboteurs destroy Kingsland ammunition depot, Lyndhurst, New Jersey.

April 1917   The U.S. declares war on Germany and enters World War I.

Bibliography

Allen, Thomas B. *Shadows in the Sea.* New York: The Lyons Press, 1996.

———. *The Shark Almanac.* New York: The Lyons Press, 1999.

Baldridge, H. D. *Shark Attack.* New York: Berkeley, 1974.

Bass, A. J., J. D. D'Aubrey, and N. Kistnasamus. "Carcharodon carcharias," *Invest. Rep. Oceanograph. Inst.* (3) Oceanographic Institute South African Associates for Marine Biological Research, Durban.

Battan, Mary. *The Shark Attack Almanac.* Los Angeles: The Kids Press, 1999.

Benchley, Peter. *Jaws.* New York: Doubleday and Company, Inc., 1974.

Bigelow, H. B. and W. C. Schroeder. *Fishes of the Western Northern Atlantic.* New Haven, Connecticut: Yale University, 1948.

Brown, Edward. "Sudden Death," *New Jersey Monthly,* pp. 45–49 (June 1986).

Brownlee, Shannon. "On the Track of the Real Shark," *Discover,* pp. 26–38 (July 1985).

Buchholz, Margaret Thomas. *Shore Chronicles.* Harvey Cedars, N.J.: Down The Shore Publishing, 1999.

Budker, P. *The Life of Sharks.* New York: Columbia University Press, 1971.

Carey, F. G., J. W. Kanwisher, O. Brazier, G. Gabrielson, J. G. Casey, and Harold Pratt, Jr. "Temperature and Activities of a White Shark, Carcharodon carcharias," *Copeia* (2) pp. 254–260 (1982).

Casey, John G. *Angler's Guide to Sharks of the Northeastern United States: Maine to Chesapeake Bay.* Washington, D.C.: Bureau of Sports Fisheries, 1964.

Casey, John G. and Harold Pratt, Jr., "Distribution of the White Shark, Carcharodon carcharias in the Western North Atlantic." *Memoirs, Southern California Academy of Science,* vol. 9, pp. 2–14 (1985).

Castro, J. I. *The Sharks of North American Waters.* College Station, Texas: Texas A&M University Press, 1983.

Clark, Eugenie. *The Lady and the Sharks.* New York: Harper & Row, 1969.

———. "Sharks: Magnificent and Misunderstood," *National Geographic,* vol. 160, no. 2, pp. 138–186 (1981).

Cliff, Geremy. "Shark Attack!" *Conserva,* November 1990: 10–13.

Collier, Ralph S., "Shark Attacks Off the California Islands: Review and Update." *Third California Islands Symposium,* pp. 453–462 (1993).

Compagno, L. J. V. *Sharks of the World.* Rome: FAO Species Catalogue. FAO Fisheries Synopsis No. 125, vol. 4, Part 1 and Part 2. United Nations Development Programme, Food and Agriculture Organization of the United Nations, 1984.

Coppleson, V. M. *Shark Attack*. London: Angus & Robertson, 1959.

Cousteau, Jacques-Yves and Phillip. *The Shark: Splendid Savage of the Sea*. Garden City, New York: Doubleday & Company, Inc., 1970.

Davies, D. H. *About Sharks and Shark Attack*. Peitermaritzburg: Shuter and Shuter, 1964.

———— and G. D. Campbell. "The Aetiology, Clinical Pathology and Treatment of Shark Attack (Based on observations in Natal, South Africa)," *Fray. Mar. Med. Serv.*, 48(3), 1962.

Dubowski, Cathy East. *Shark Attack*. New York: DK Publishing, 1998.

Ellis, Richard. *Book of Sharks*. New York: Grossett & Dunlap, 1975.

———— and John McCosker. "Speaking of Sharks," *Oceans,* vol. 19, pp. 24–29, 58 (June 1986).

———— and J. McCosker. *Great White Shark*. New York: HarperCollins, in collaboration with Stanford University Press, 1991.

Fernicola, Richard G. *In Search of the Jersey Man-Eater*. Deal, N.J.: George Marine Library, 1987.

————. *Tracking the Jersey Man-Eater*. VHS Documentary, Deal, N.J.: George Marine Library, 1990.

Fowler, Henry W. "Lampreys and Sharks of New Jersey," *Bulletin 7. N.J. State Museum* (June 1959).

Freuchen, Peter and David Loth. *Book of the Seven Seas*. New York: Julian Messner, Inc., 1957.

Geiser, John. *The Shore Catch*. Asbury Park, New Jersey: Asbury Park Press, Inc., 1984.

Gilbert, P.W. (ed.). *Sharks and Survival*. Boston: D. C. Heath and Company, 1963.

————— Mathewson, and D. P. Rall (eds). *Sharks, Skates, and Rays*. Baltimore: Johns Hopkins Press, 1967.

Gosner, Kenneth L. *Guide to Identification of Marine and Estuarine Invertebrates*. New York: Wiley-Interscience, 1971.

Greenberg, Idaz, Jerry Greenberg, and Michael Greenberg. *Sharks and Other Dangerous Sea Creatures*. Miami: Seahawk Press, 1981.

Gudger, E. W. "A Boy Attacked by a Shark, July 25, 1936 in Buzzards Bay, Massachusetts," *The American Midland Naturalist*, vol. 44, pp. 714–719 (1950).

Guyton, Arthur C. *Human Physiology and Mechanisms of Disease*. Philadelphia: W.B. Saunders, 1982.

Hellman, Geoffrey. *Bankers, Bones, and Beetles*. Garden City, New York: Natural History Press, 1969.

Helm, Thomas. *Shark! Unpredictable Killer of the Sea*. New York: Dodd, Mead and Co., 1962.

Griffis, Nixon. *The Mariner's Guide to Oceanography*. New York: Hearst Maine Books, 1985.

Gruber, S.H. (ed.). *Discovering Sharks*. Highlands, New Jersey: American Littoral Society, 1991.

Hodgson, E. S. and R. F. Mathewson (eds.). *Sensory Biology of Sharks, Skates, and Rays*. Arlington, Virginia: Office of Naval Research Dept., 1978.

Hutchins, B. "Megamouth: gentle giant of the deep." *Australian Natural History*, 23 (12):9100–17 (1992).

Jersey Coast Shark Anglers. "Shark Chart: Catches According to Water Temperature," 22nd Annual Shark Tournament program, p. 2 (June 24, 2000).

Kandel, Eric R. and James H. Schwartz. *Principles of Neural Science.* New York, Elsevier/North Holland, 1983.

Kenney, Nathaniel T., "Sharks: Wolves of the Sea." *National Geographic,* vol. 133, No. 2., pp. 222–257.

Klimley, A. Peter. "The Predatory Behavior of the White Shark." *American Scientist,* vol. 82, March–April 1994, pp. 123–33.

Llano, George A. *Sharks: Attacks on Man.* New York: Grossett & Dunlap, 1975.

Lineaweaver, T. H. and R. H. Backus. *The Natural History of Sharks.* New York: The Lyons Press, 1984.

Lloyd, John Bailey. *Eighteen Miles on History.* Surf City, N.J.: Sandpiper Inc., 1986.

Macleish, W. H. (ed.). "Sharks," *Oceanus* 24 (4):1–79 (1981).

McClane, A. J. *McClane's New Standard Fishing Encyclopedia.* New York: Gramercy Books, 1998.

Mclaughlin, Bill. "Sharks Have Become Popular Quarry," *Asbury Park Press,* August 16, 1985.

Methot, June. *Up & Down the Beach.* Navesink, N.J.: Whip Publisher, 1988.

Miller, D. I. and R. S. Collier. "Shark Attacks in California and Oregon, 1926–1979," *California Fish and Game,* 67:76–104 (1980).

Moss, George H. and Karen Schitzpan. *Those Innocent Years.* Navesink, N.J.: Ploughshire Press, 1996.

Moss, S. A. *Sharks: An Introduction for the Amateur Naturalist.* Englewood Cliffs, New Jersey: Prentice-Hall, 1984.

National Marine Fisheries Service, "Fishery Management Plan for Sharks of the Atlantic Ocean." National Oceanic and Atmospheric Administration, U.S. Department of Commerce.

Nichols, J. T. and R. C. Murphy. "Long Island Fauna. IV. The Sharks (Order Selachii)" and "The Shark Situation in the Waters About New York" *Brooklyn Museum Science Bulletin,* vol. 3 no. 1 (April 1916) and vol. III, no. 4 (October 1916).

Noyes, G. Harold. "Climatological Data: New Jersey Section" vol. XXIX, no. 6 and no. 7, U.S. Department of Agriculture, Weather Bureau (June–July 1916).

Perrine, Doug. *Sharks & Rays of the World.* Stillwater, Minnesota: Voyageur Press, 1999.

Pratt, H. L., Jr., J. G. Casey, and R. B. Conklin. "Observations on Large White Sharks, Carcharodon carcharias, Off Long Island, New York." *Fishery Bulletin:* vol. 80, no. 1, pp. 153–156 (1982).

Proceedings of the Board of Trustees. *Ninth Annual Report of the Board of Regents of the Smithsonian Institution, Showing the Operations, Expenditures, and Condition of the Institution up to January 1, 1855.* Washington, D.C.: Beverly Tucker, Senate Printer, 1855.

Reader's Digest (eds.) *Secrets of the Seas.* Pleasantville, New York: Reader's Digest Association, 1972.

——— (eds.) *Sharks, Silent Hunters of the Deep.* New York: Reader's Digest Services, 1986.

Ristori, Al., "Blue Shark Comeback," *Saltwater Sportsman,* pp. 56–60 (January 1986).

Schaefer, Henry. "Fishing Captain Recalls the Day a Shark Killed Two in Matawan," *The Daily Register,* Red Bank, N.J., July 23 (1975).

Schaeffer, J. Parsons (ed.). *Morris' Human Anatomy.* Philadelphia: The Blakiston Company, 1942.

Scientific American. "Sharks, Man-Eating and Otherwise—The Present Status of a Very Old Controversy" (July 24, 1916).

Sibley, G., J. A. Seigel, and C. C. Swift (eds.). "Biology of the White Shark." *Memoirs of the Southern California Acdemy of Sciences,* 9:1–150 (1985).

Steel, R. *Sharks of the World.* New York: Blandford Press, 1985.

Stevens, J. D. (ed.). *Sharks.* New York: Facts on File, 1987.

————. "Sharks." *Oceanus,* 24 (4) Winter 1981/82.

Wallett, Timothy. *Shark Attack and Treatment of Victims in South African Waters.* Capetown: Macdonal Purnell, 1978.

Watson, John F. *Annals of the Philadelphia and Pennsylvania, In The Olden Time,* vol. II. Philadelphia: Elijah Thomas, 1860.

Wexler, Mark. "Facing Up to Our Fears about Sharks," *National Wildlife,* vol. 20, Number 5, pp. 4–10 (Aug.–Sept. 1982).

Wilson, E. O., "In Praise of Sharks," *Discover,* pp. 40–53 (July 1985).

Witcover, Jules. *Sabotage at Black Tom Island: Imperial Germany's Secret War in America, 1914–1917.* Chapel Hill, N.C.: Algonquin, 1989.

Wood, Don. *The Unique New York and Long Branch.* New York: Hillcrest Press, 1985.

Young, Peter. *World Book Encyclopedia.* Field Enterprise Education Corp., vol. 21, 1972.

Young, W. E., and H. S. Mazet. *Shark! Shark.* New York: Gotham House, 1933.

Younghusband, Peter. "Tourists Breathe Easy on the Beaches of Durban, South Africa, but It Wasn't Always That Way," *International Wildlife,* pp. 4–10 (Jan.–Feb. 1982).

Zahuranec, B. J. *Shark Repellants from the Sea-New Perspectives.* American Association for Advancement of Science, Washington, D.C.: AAAS Selected Symposium 83, 1983.

NEWSPAPERS
The Asbury Park Press
The Atlantic City Daily Press
The Baltimore Sun
The Boston Globe
The Bronx Home-News
The Brooklyn Reporter
The Chicago Tribune
The Chicago Sun Times
The Coast Star (Manasquan, N.J.)
The Courier News (Beach Haven, N.J.)
The Daily Register (Shrewsbury, N.J.)
The Evening Bulletin (Philadelphia)
The London Times
The Matawan Journal
The Monmouth Democrat
The New Brunswick Times (N.J.)
The Newark Star-Eagle
The New York Daily News
The New York American
The New York Herald
The New York Sun
The *New York Times*
The News-Journal (Daytona Beach)
The Ocean Grove Times
The Public Ledger (Philadelphia)
The Philadelphia Inquirer
The San Francisco Chronicles
The Washington Herald
The Washington Post

Index